Geography Skills for NCEA Level Three

2nd Edition

Justin Peat

Australia • Brazil • Mexico • Singapore • United Kingdom • United States

Geography Skills for NCEA Level Three
2nd Edition
Justin Peat

Cover design: Macarn Design
Text design: Smartwork Creative Ltd
Production controller: Siew Han Ong

ACKNOWLEDGEMENTS
The authors and publisher wish to thank the following people and organisations for permission to use the resources in this textbook. Every effort has been made to trace and acknowledge all copyright owners of material used.

Pages 7, 94, and 160 courtesy of Land Information New Zealand (LINZ); front cover, pages 7 (bottom) 9, 30 (bottom), 43, 54 (top right), 67 (bottom), 68 (bottom left and right), 72, 81 (bottom left and right), 96 (top left and right), 110 (top left and right), 133 and 145 courtesy of Shutterstock; page 30 courtesy of GeoEye; page 40 (bottom) courtesy of GNIS; page 54 (bottom) courtesy of NOAA; page 55 courtesy of Mark Elstone; page 56 courtesy of New Zealand Herald; pages 58, 94 and 97 courtesy of Fairfax Media; page 98 courtesy of Forest and Bird; page 110 (bottom) courtesy of the Green Party; pages 13, 14, 70 and 143 courtesy of NASA; page 146 (bottom) courtesy of NSIDC; page 148 courtesy of NOAA; pages 162-63 courtesy of New Zealand Wind Energy Association.

For product information and technology assistance,
in Australia call **1300 790 853**;
in New Zealand call **0800 449 725**

For permission to use material from this text or product, please email **aust.permissions@cengage.com**

National Library of New Zealand Cataloguing-in-Publication Data
A catalogue record for this book is available from the National Library of New Zealand.

978 017 042528 5

Cengage Learning Australia
Level 7, 80 Dorcas Street
South Melbourne, Victoria Australia 3205

Cengage Learning New Zealand
Unit 4B Rosedale Office Park
331 Rosedale Road, Albany, North Shore 0632, NZ

For learning solutions, visit **cengage.co.nz**

Printed in Malaysia by Papercraft.
6 7 24

Contents

Geographic concepts and skills

The aim of Achievement Standard Geography 3.4 is to test your *understanding of a given environment through selection and application of geographic concepts and skills.* In essence, the standard assesses your ability to:

1 analyse an environment through selection and application of geographic concepts; and

2 select and use geographic skills and conventions in the presentation and/or interpretation of information.

Geography 3.4 builds on the skills introduced and developed in Levels 1 and 2 Geography. However, it does place more emphasis on the application of geographic concepts to demonstrate understanding than it does the selection and usage of geographic skills. Moreover, such is the increased importance of geographic concepts at Level 3, their selection and application are assessed to Excellence level, whereas the selection and usage of geographic skills and conventions, and interpretation of information is only assessed to Merit level (Figure 1).

	Criteria	Explanatory notes
Achieved	Demonstrate **understanding** of a given environment(s) through selection and application of geographic concepts and skills involves:	• Analysing the environment(s) through selection and application of geographic concepts. • Selecting and using geographic skills and conventions in the presentation and/or interpretation of information.
Merit	Demonstrate **in-depth understanding** of a given environment(s) through selection and application of geographic concepts and skills involves:	• Analysing the environment(s) **in detail** through selection and application of geographic concepts • Selecting and using geographic skills and conventions **with precision** in the presentation and/or interpretation of information.
Excellence	Demonstrate **comprehensive understanding** of a given environment(s) through selection and application of geographic concepts and skills involves:	• Analysing the environment(s) **with insight** through selection and application of geographic concepts.

Figure 1 Achievement criteria for Geography 3.4

It follows then, that to demonstrate a comprehensive understanding of an environment through the selection and application of geographic concepts, you should refer to the requisite concepts when responding to the learning activities in this book and in the final examination.

ISBN: 9780170425285

Revisiting the geographic concepts

The *Teaching and Learning Guide for Geography* identifies seven key concepts that serve to 'provide a framework that geographers use to interpret and represent information about the world' (Figure 2). The list, however, is not exhaustive and you can choose to utilise other geographic concepts so long as the concept has a spatial component, e.g. globalisation, location and accessibility.

- Globalisation: The process of increasing interaction between people and ideas on a global scale due to advances in technology, communication and transportation.
- Location: The position of something that can be given in absolute terms, or in relation to other objects.
- Accessibility: A measure of the ease of movement of people or ideas.

Several Maori concepts also have relevance to Geography including kaitiakitanga (to care for), taonga (physical or cultural resource) and hekenga (migration).

Environments

Environments may be natural and/or cultural. They have particular characteristics and features which can be the result of natural and/or cultural processes. The particular characteristics of an environment may be similar to and/or different from another.

Perspectives

Perspectives are ways of seeing the world that help explain differences in decisions about, responses to, and interactions with environments. Perspectives are bodies of thought, theories or world-views that shape people's values and have built up over time. They involve people's perceptions (how they view and interpret environments) and viewpoints (what they think) about geographic issues. Perceptions and viewpoints are influenced by people's values (deeply held beliefs about what is important or desirable).

Processes

A process is a sequence of actions, natural and/or cultural, that shape and change environments, places and societies. Some examples of geographic processes include erosion, migration, desertification and globalisation.

Patterns

Patterns may be spatial: the arrangement of features on the earth's surface; or temporal: how characteristics differ over time in recognisable ways.

Interaction

Interaction involves elements of an environment affecting each other and being linked together. Interaction incorporates movement, flows, connections, links and interrelationships. Landscapes are the visible outcome of interactions. Interaction can bring about environmental change.

Change

Change involves any alteration to the natural or cultural environment. Change can be spatial and/or temporal. Change is a normal process in both natural and cultural environments. It occurs at varying rates, at different times and in different places. Some changes are predictable, recurrent or cyclic, while others are unpredictable or erratic. Change can bring about further change.

Sustainability

Sustainability involves adopting ways of thinking and behaving that allow individuals, groups and societies to meet their needs and aspirations without preventing future generations from meeting theirs. Sustainable interaction with the environment may be achieved by preventing, limiting, minimising or correcting environmental damage to water, air and soil, as well as considering ecosystems and problems related to waste, noise, and visual pollution.

Figure 2 The prescribed geographic concepts

ISBN: 9780170425285

Selecting and using geographic skills and conventions

Geographic skills are assessed in the external examination standard Geography 3.4 and the internal research standard Geography 3.5.

In the external examination you will be provided with a resource booklet, which you will use to show your understanding and application of geographic skills and concepts. The booklet will include a variety of resources such as maps, tables, diagrams, photographs and opinions. The resources in the booklet will be unfamiliar to you but will generally be about a particular geographic issue, which could be from New Zealand or an overseas setting.

Basic geographic skills

Visual
- Interpretation of photographs, cartoons or diagrams including age-sex pyramids and models such as a wind rose
- Interpreting and completing a continuum to show value positions

Mapping
- Use of six-figure grid references and latitude and longitude
- Compass direction and bearings
- Distance, scale, area calculation
- Location of natural and cultural features
- Determination of height, cross-sections
- Use of a key, précis map construction
- Recognition of relationships, application of concepts
- Interpretation of other geographic maps like weather maps, cartograms, choropleth maps

Graphing
- Interpretation and construction of bar graphs (single and multiple), line graphs (single and multiple), pie and percentage bar graphs, scatter graphs, dot distribution, pictograms and climate graphs

Tables
- Recognition of pattern
- Simple calculation such as mean, mode and conversion to percentages

Figure 3 Basic geographic skills

No new skills are introduced at Level 3. However, the complexity of the material in the examination resource booklet will be higher than that provided at Levels 1 and 2 (Figure 4). This is because the achievement standard at Level 3 intends to assess your ability to select and apply skills with accuracy from the resource material presented, rather than simply present or interpret data as was required at Levels 1 and 2. For example, you may be asked to determine the location of a natural feature requiring the use of a range of resources and a combination of location skills such as six-figure grid references, latitude and longitude, or distance and bearing. Alternatively, you could be given a table of data and asked to select and construct an appropriate graph to illustrate the data.

	Basic skills at NCEA Level 1	Complex skills at NCEA Level 2 and Level 3
Latitude and longitude	• Degrees and minutes only.	• Degrees, minutes and seconds.
Direction	• To nearest inter-cardinal point (8-point compass).	• To the nearest 16-point compass direction.
Scale	• Simple linear scale measurement on a map. • Recognition of different scales.	• Changed ratio scale with size. • Converting linear to ratio or vice versa. • Use of other scales apart from distance, i.e. time.
Graphing	• Scatter graph (interpretation or completion only).	• Triangular graphs • Cumulative graphs • Scatter graph construction • Positive/negative • Statistical mapping • Proportional circle maps
Tables	• All statistics given are used.	• Not all statistics given may be necessary for completion of tasks. • Percentage change calculations.

Figure 4 Basic and complex skills

ISBN: 9780170425285

Mapping skills

As geographers, knowing how to interpret a map correctly is an essential skill, as it provides us with information about places and helps us to identify patterns and changes in a landscape. The most common map used by geographers is the topographic map (Figure 12). Topographic maps are detailed, accurate graphic presentations of features on the earth's surface. Examples of features shown on topographic maps include:

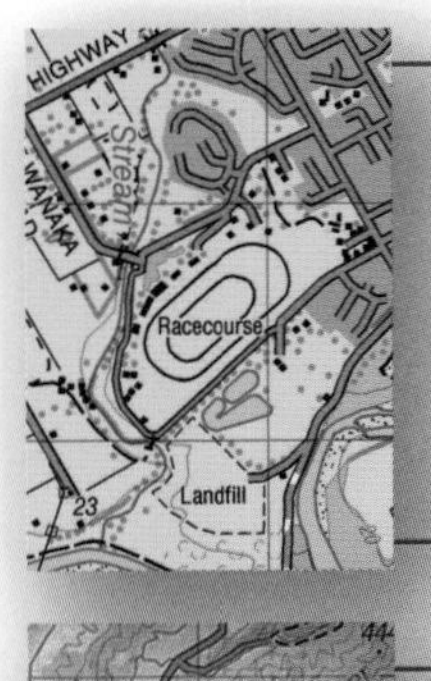

Cultural features

- roads
- buildings, urban areas
- transport networks
- place names

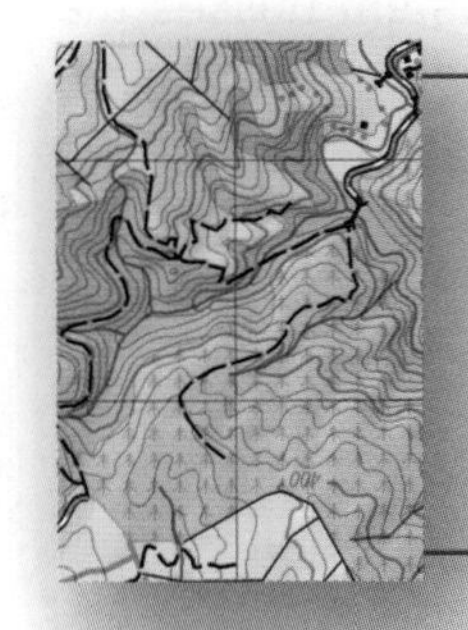

Vegetation features

- native and plantation forest
- vineyards and orchards
- scrub

Relief features

- mountains
- cliffs
- ridges and valleys
- slopes and depressions

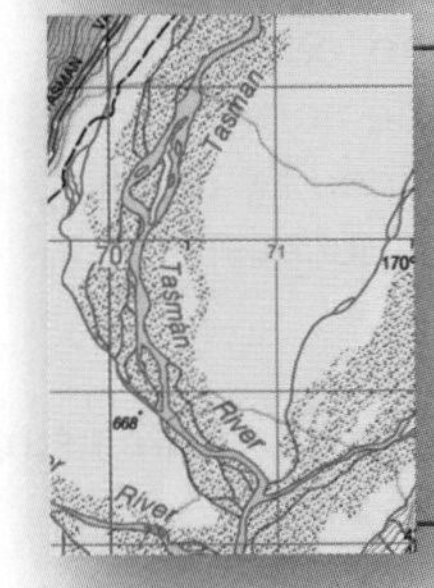

Drainage features

- rivers
- lakes
- streams
- swamps

Figure 5

Location, direction and bearing

The notion of location (or the particular place where something is) is the most basic of all geographical concepts. It is therefore important that as a student of geography you refine the skill of locating places before mastering any other.

Location is usually expressed in one of two ways:

- **Relative location** describes the location of a place or feature as it relates to other features.
- **Absolute location** refers to the exact location of a point on the earth's surface, as it appears on a map referenced to a grid system. Of the many grid-referencing systems in use, the six-figure grid reference system, and latitude and longitude are used most.

Figure 6 Sixteen-point compass and associated bearing

Relative location is usually expressed in one of two ways. The most common method is to describe direction according to the points of a compass; however, it can also be expressed as a bearing when a more precise reading is required. In Level 3 you need to know the four cardinal (main) points of the compass: north (N), south (S), east (E) and west (W); the four intermediate points: northeast (NE), southeast (SE), southwest (SW) and northwest (NW); and the additional eight points that form the 16-point compass rose, e.g. north-northeast (NNE), east-northeast (ENE), etc. (Figure 6).

Direction can also be expressed as an angle or bearing. On a map, the bearing between two locations can be measured with the aid of a protractor. Bearings are always measured clockwise from north (0°). Using this principle, the angle 90° clockwise of north is east and southeast is located at an angle of 135° clockwise of north. In Figure 7, Landmark Feature A is located at an angle of 320° clockwise of north, while Landmark Feature B is located at an angle of 40° clockwise of north.

ISBN: 9780170425285

Grid references

The absolute location of a feature on a map can be found through the use of a grid system. Six-figure reference systems are regularly used to establish the exact location of features on New Zealand's topographic 1:50,000 map series the *Topo50*.

The six-figure grid reference system utilises the equally spaced vertical and horizontal grid lines found on a topographical map to determine a feature's absolute location. Referred to as **eastings** and **northings**, these intersect each other on a map to form grid squares. The grid squares are then used to help calculate unique six-figure grid references (GR) for individual features (Figure 8). Six-figure grid references contain six digits. The first three digits of a six-figure grid reference refer to the easting, and the second three digits refer to the northing. The third and sixth digits are determined by dividing the easting and northing, respectively, into tenths (Figure 9).

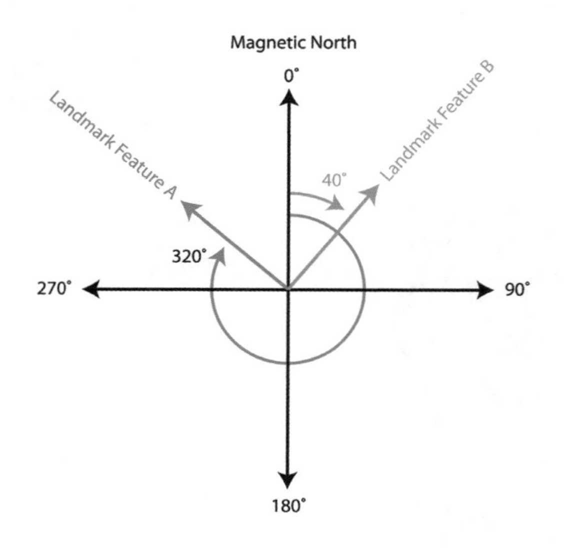

Figure 7 Measuring bearing

To identify the absolute location of a feature using six-figure grid reference system for the example below (Figure 9), follow the steps below:

1. To determine the six-figure grid reference for the symbol in Figure 9 identify the first vertical grid line to the left of the symbol. In the example below, the first grid line to the left is labelled 23 in the bottom margin. This is the easting.
2. Estimate the number of tenths eastward the symbol is from the easting on the left. In the example below, the symbol is located a distance of *five-tenths* from the easting on the left. This gives a final easting reading of 235.
3. Now identify the first horizontal grid line below the symbol. In the example below, the first grid line below is labelled 10 in the left margin. This is the northing.
4. Estimate the number of tenths northward the symbol is from the northing beneath. In the example above, the symbol is located a distance of *seven-tenths* from the northing beneath. This gives a final northing reading of 107.
5. You have now calculated the six-figure grid reference for the symbol. It should be written as GR235107.

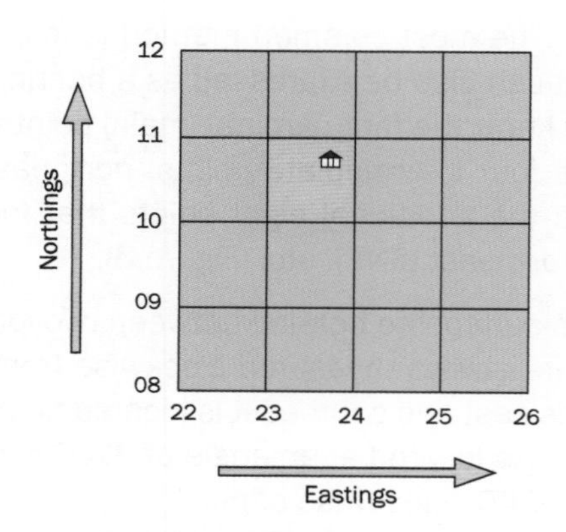

Figure 8 Northing and eastings

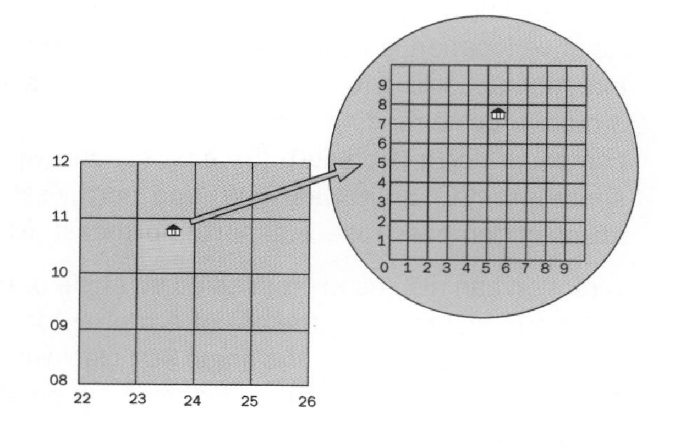

Figure 9 Determining six-figure references

ISBN: 9780170425285

Latitude and longitude

Small-scale maps depicting a country or continent use latitude and longitude to determine absolute location (Figure 10).

Lines of **latitude** run east to west around the earth's circumference. They run parallel to each other and for this reason are also known as parallels of latitude. Latitude is measured in degrees north (N) or south (S) in relation to the equator, which itself represents zero degrees (0°). The equator divides the earth into the northern hemisphere and southern hemisphere. The latitude of the North Pole is 90°N of the equator while the latitude of the South Pole is 90°S of the equator.

Lines of **longitude** run in a north to south direction from the North Pole to the South Pole, and on a two-dimensional map intersect lines of latitude at right angles. Longitude is measured in degrees west (W) or east (E) of the prime meridian (0°). The prime meridian divides the earth into the western hemisphere and eastern hemisphere.

Each degree can be divided into smaller units called minutes (') and each minute of each degree can be divided into seconds ("). There are 60 minutes in each degree of latitude or longitude and 60 seconds in every minute.

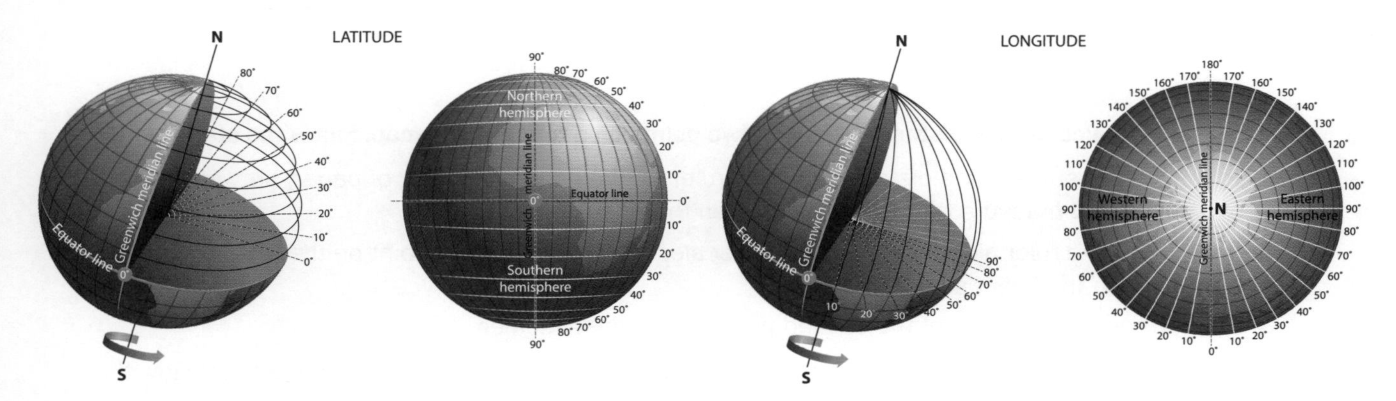

Figure 10 Latitude and longitude

To identify the exact location of a feature using latitude and longitude, follow the steps below:

1. Using a world or country map, locate the feature you wish to calculate latitude and longitude for.
2. Having found a place or feature, identify the line of latitude directly to the south if the feature is in the northern hemisphere or directly to the north of the feature if the feature is in the southern hemisphere. This is your line of latitude in degrees (°) north (N) or south (S) of the equator.
3. Now imagine that the space between each line of latitude is divided into 60 minutes. Estimate how many sixtieths from the line of latitude the feature is located. This is the number of minutes (') in addition to the degrees north or south of the equator your line of latitude is located.
4. Next, imagine that the space between each minute of latitude is divided into 60 seconds. Estimate how many sixtieths between the two readings the feature is located. This is the number of seconds (") in addition to the degrees and minutes north or south of the equator your feature is located.
5. Repeat the process above to identify the line of longitude. If your feature is located in the eastern hemisphere, identify the line of longitude directly to the west. If it is in the western hemisphere, identify the line of longitude directly to the east. This is your line of longitude in degrees (°) west (W) or east (E) of the prime meridian.
6. Now estimate the number of minutes and seconds as above.
7. Remember to always state latitude first and longitude second.

ISBN: 9780170425285

Using scale to calculate distance

A map represents an area on the earth's surface and a map's scale refers to the relationship between distances portrayed on a map and the distance on the ground. Put in another way, a map is a scaled-down representation of part of the earth's surface. This means that a map's scale can be used to calculate surface distances represented by the map.

There are three accepted ways of showing scale on a topographic map:

- As a **written statement**. For example, on the New Zealand *Topo50* series, 1 cm is equal to 50,000 cm (or 500 metres).
- As a **ratio or representative fraction**. For example, the 1:50,000 ratio of the *Topo50* series can be expressed as the fraction 1/50000. Here, the numerator represents the number of units on the map while the denominator represents the number of units that one unit on the map is equal to on the ground. In this example, 1/50000 means that one unit on the map equals 50,000 units on the ground.
- As a **linear scale** or scale bar. For example:

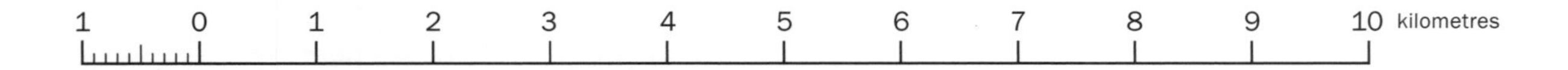

To use scale to calculate the distance between two points on a topographic map, follow the steps below:

1. To measure a straight-line distance, place a ruler (or the edge of a sheet of paper) between the two points and measure the distance between them.
2. Next, place the ruler along the map's linear scale, overlaying the zero point on the ruler with the zero point on the linear scale.
3. Finally, read the distance of the second point off the linear scale.

To estimate distance along a curve (e.g. a river), replace the ruler with a piece of string and follow the steps above.

Using scale to estimate area

The area that a feature on the earth's surface covers can be calculated by using the scale of the map.

To estimate an area from a topographic map, follow the steps below:

1. Measure the distance of the map area from east to west in kilometres.
2. Measure the distance of the map area from north to south in kilometres.
3. If the area to be measured is square or rectangular, multiply the east–west distance by the north–south distance.
4. If the area to be measured is an irregular shape, an accurate calculation will be difficult. If the area to be measured is an irregular shape, an accurate calculation will be difficult. Instead, estimate the area by counting the number of squares covered by more than half of the feature and ignoring squares covered by less than half of the feature.
5. Express the area in square kilometres (km^2).

For example, the area covered by the lake in Figure 11 is 58 km^2.

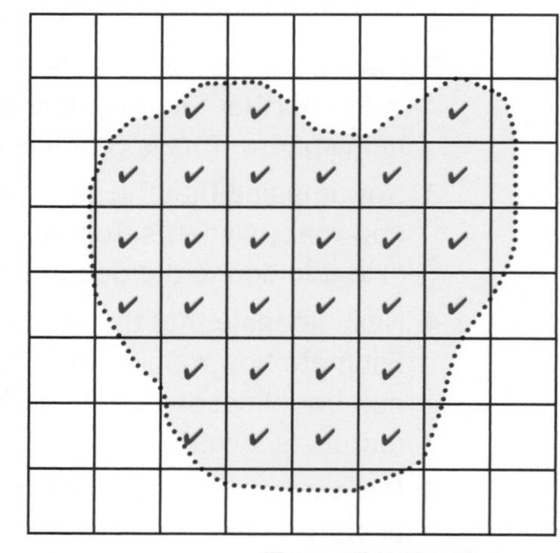

Figure 11 Estimating area

☐ = 2 km^2 ✔ = 29 Area = 58 km^2

ISBN: 9780170425285

Interpreting maps with PQE

There are a number of methods used by geographers to interpret data presented in maps. The PQE method is particularly useful when describing patterns in data. The letters **PQE** stand for: **P**attern, **Q**uantify and **E**xceptions.

- **Pattern:** To describe the various patterns on a map, look for elements or features of the map that stand out, or concentrations of a particular colour or feature.
- **Quantify:** Quantifying requires you to extract specific and accurate detail from the map to support the patterns you have identified. This involves using the statistics, depths, spot heights and locations to give specific details.
- **Exceptions:** Often you will notice elements or features in the data that do not fit the overall pattern you have identified. These are the exceptions that also need to be identified and quantified.

Interpreting relief

Relief is a term used by geographers to describe the shape or pattern of landforms, its height and the steepness of its slopes.

Topographic maps show the pattern of relief in three ways:

- shading
- spot heights
- contour lines.

Of the three methods used, contour lines are most effective in showing relief patterns. Contour lines trace out areas of equal height or elevation above sea level, and give an indication of the gradient or slope of the land. Closely spaced contour lines indicate a steep gradient and contour lines spaced far apart indicate a gentle gradient (Figure 13).

The vertical distance between adjacent contour lines is known as the **contour interval** (CI).

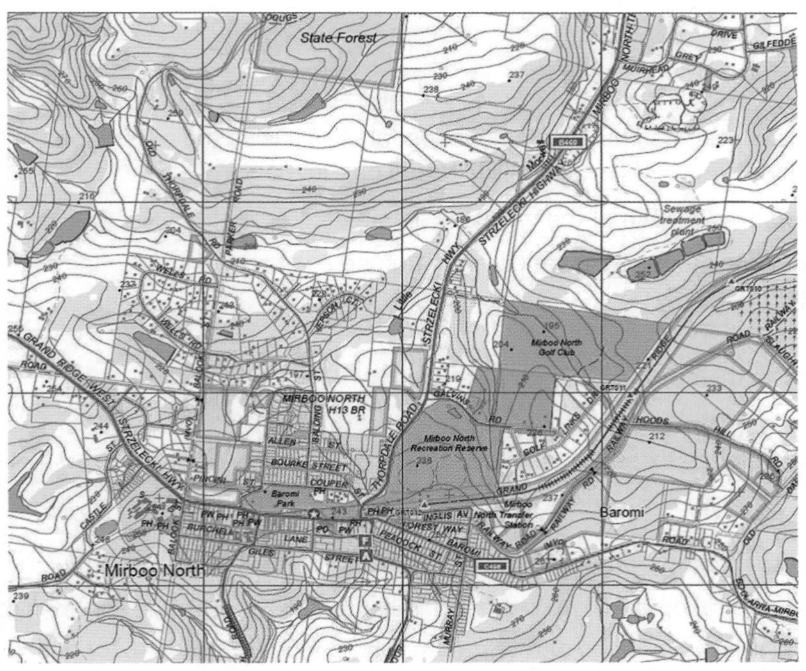

Figure 12 Topographic map of Mirboo North, Victoria, Australia

To use contour lines to determine the elevation of a feature on a topographic map, follow the steps below:

1. Find the contour interval of the map from the key or legend, and note both the interval and the unit of measure. New Zealand's *Topo50* map series, for example, has a contour interval of 20 metres.
2. Find the numbered contour line nearest the feature for which the elevation is being sought.
3. Determine the direction of slope from the numbered contour line to the desired feature.

ISBN: 9780170425285

4 Count the number of contour lines that must be crossed to go from the numbered line to the feature and note the direction up or down. The number of lines crossed multiplied by the contour interval is the vertical distance above or below the starting value.

5 When the feature is on a contour line, its elevation is that of the contour. If the feature is between contour lines, then estimate the elevation to be one-half of the contour interval.

Common contour patterns

A skilled geographer can visualise the shape of the landforms by studying patterns created by contour lines.

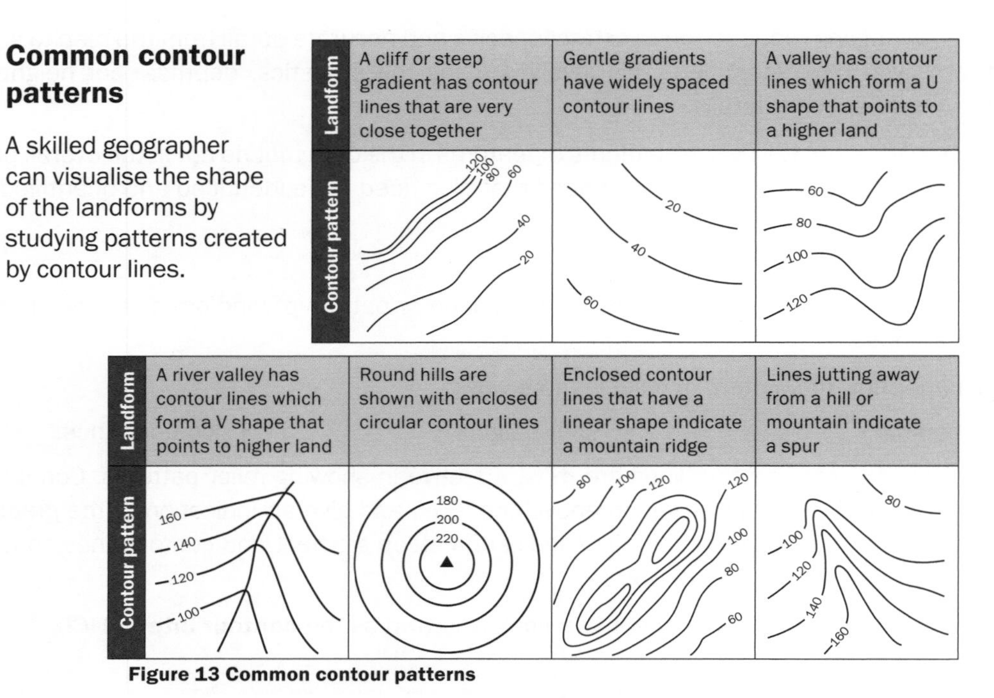

Figure 13 Common contour patterns

Précis maps

Topographic maps contain an immense amount of detail so it is sometimes helpful to construct a précis map to highlight just a few of the features illustrated on the map.

A précis map should have each of the following elements:

- a title that defines the map's location and purpose e.g. Land use surrounding the Lakes Waikare and Rotongaro.
- orientation (north point)
- legend (a key to the symbols and colours used)
- scale (linear or ratio)
- a border to define the extent of the map.

To construct a précis map, follow the steps below:

1 Identify from the topographic map the particular feature or features you wish to study.

2 Establish the relative location of the feature on the topographic map (or absolute location if the use of a six-figure grid reference is required).

3 Draw a simple outline of each feature on your précis map, taking care to place features in their correct location relative to others. You may choose to draw a grid on your précis that corresponds to the grid on the topographic map to ensure even greater accuracy. Shade in each feature in an appropriate colour, for example, blue for water features, black for cultural features and green for vegetation.

4 Summarise important features by constructing a key or legend to show the meaning of any symbols or shading used.

ISBN: 9780170425285

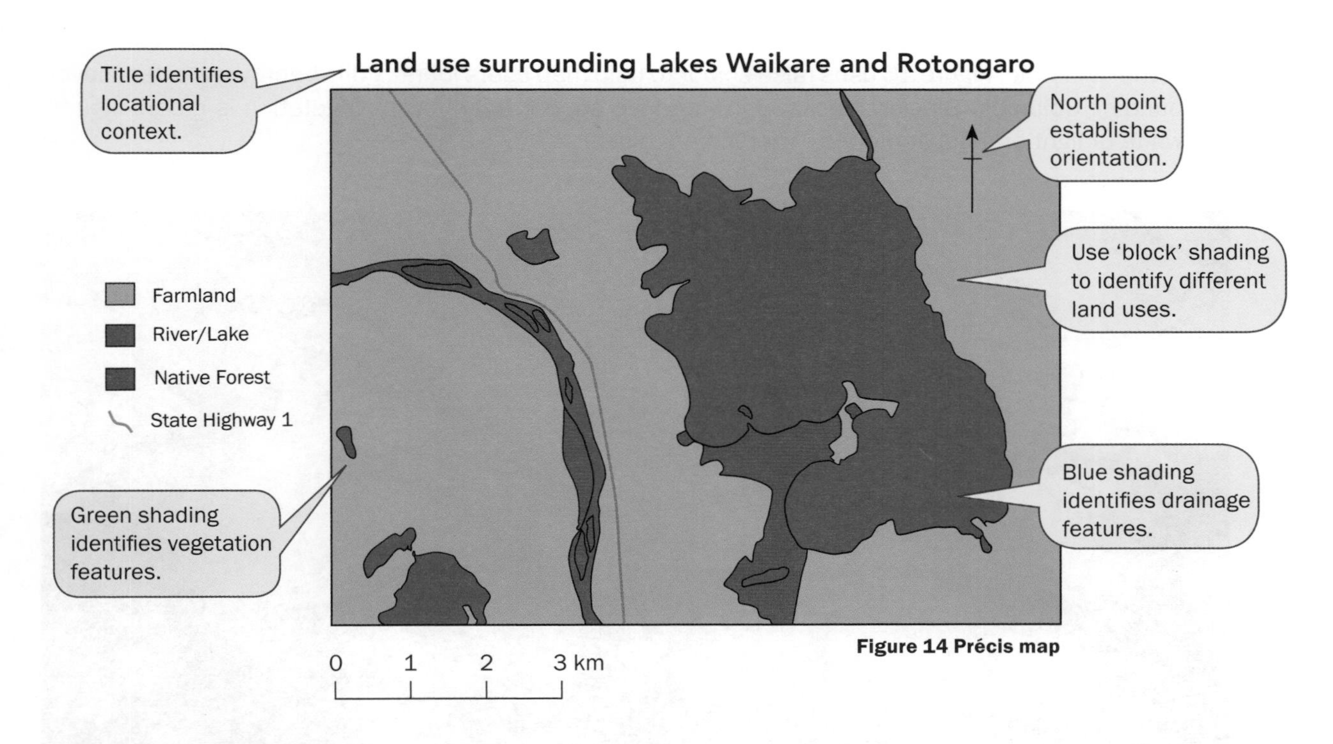

Figure 14 Précis map

Thematic maps

Thematic maps show the distribution (i.e. concentration, dispersal or flow) of geographic phenomena. They have particular geographic themes and are usually produced for specific audiences. There are many different types of thematic map, each differing according to its use and purpose.

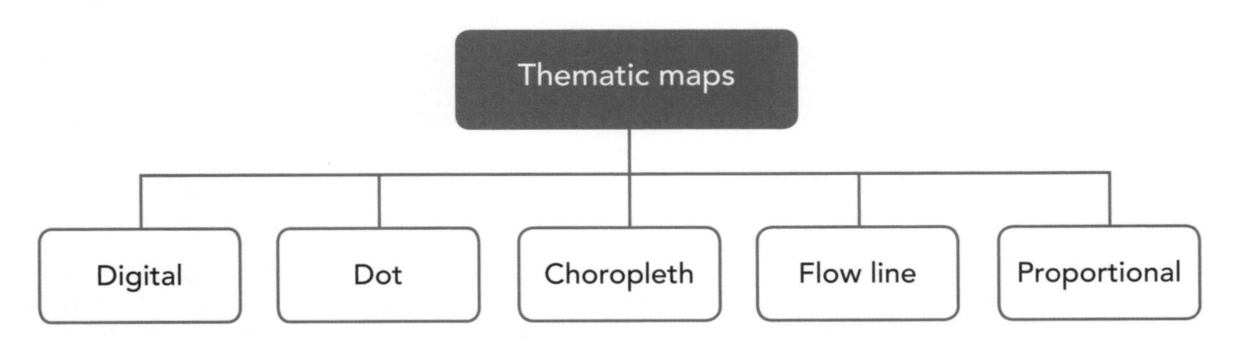

Digital maps

At any moment in time, there are more than 4600 man-made satellites orbiting the Earth. The majority of satellites are used to facilitate communications; however, many are used to observe the Earth and collect data about the height and shape of the land. Cartographers use computer programs to interpret the data to generate digital maps that highlight different aspects of the environment. For example, Figure 15 uses data collected by an orbiting NASA satellite to create a digital terrain model of the southern aspect of Mt Everest and the surrounding area.

Figure 15 A digital terrain model of Mt Everest's south face

ISBN: 9780170425285

In another example, Figure 16 uses false colour to help map users identify different types of landcover surrounding Columbia Glacier, Alaska. Snow and ice appear bright cyan, vegetation is green, clouds are white or light orange, and open water is dark blue.

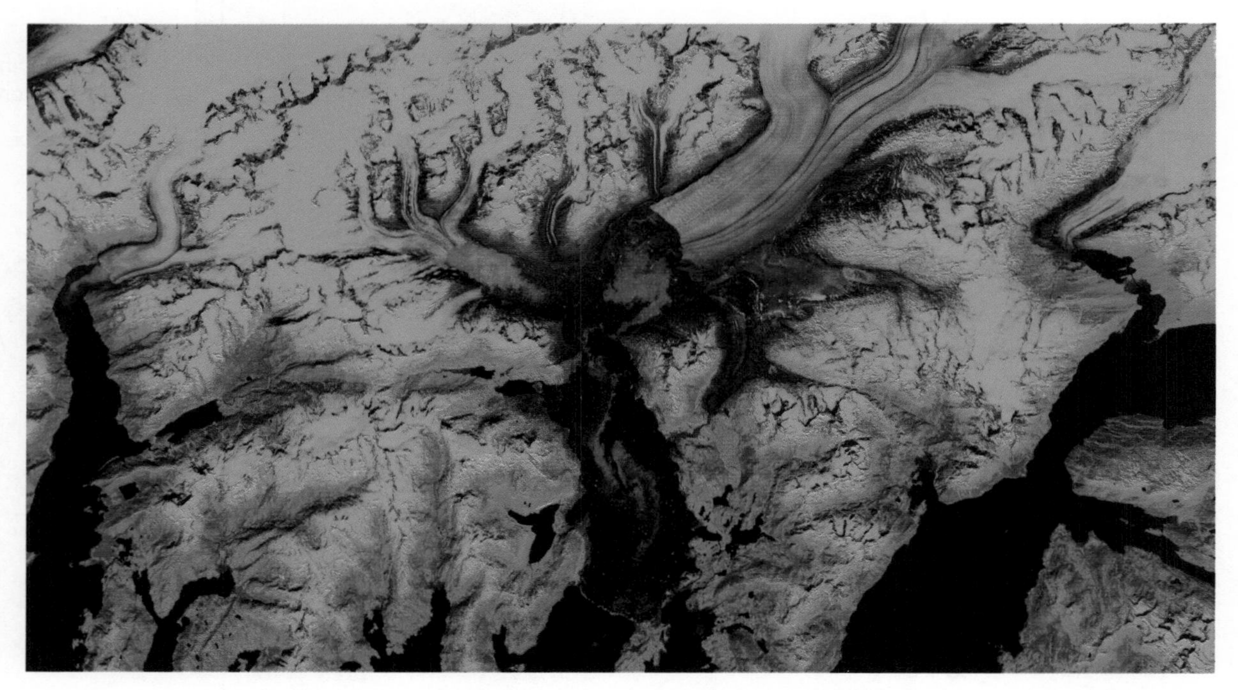

Figure 16 A digital map of Columbia Glacier, Alaska, USA

Dot maps

A dot map uses dots to show the distribution (or relative density) of geographic phenomena between different regions on a map. Each dot on a dot map represents a single feature or, in most cases, a quantity or number value. When viewed as a whole, a dot map will give its reader an impression of the overall distribution (or spread) of geographic phenomena throughout the area being mapped.

Dot maps have a wide variety of uses. For example, in medical geography, dot maps have been used to show the spread of disease in a rural community, while in physical geography, they have been used to show the pattern of earthquakes across a region. However, dot maps are most commonly used to compare population densities across regions. A variation on the standard dot map is the multi-variable dot map. Multi-variable dot maps employ different colours to compare the distribution of related phenomena or sub-groups within the same phenomena. The dot map in Figure 17 shows the degree of racial segregation (separation) in New Orleans, USA.

To interpret a dot map, follow the steps below:

1. Identify the geographic feature or phenomena being mapped.
2. Verify the dot value. Reading the map's legend can help you do this.
3. Identify the scale of the administrative regions shown on the map (i.e. does the map show neighbourhoods, census areas, states or countries?).
4. Calculate the total value of features in each area of the map.
5. Describe the distribution of the feature both within and between different areas of the map.

ISBN: 9780170425285

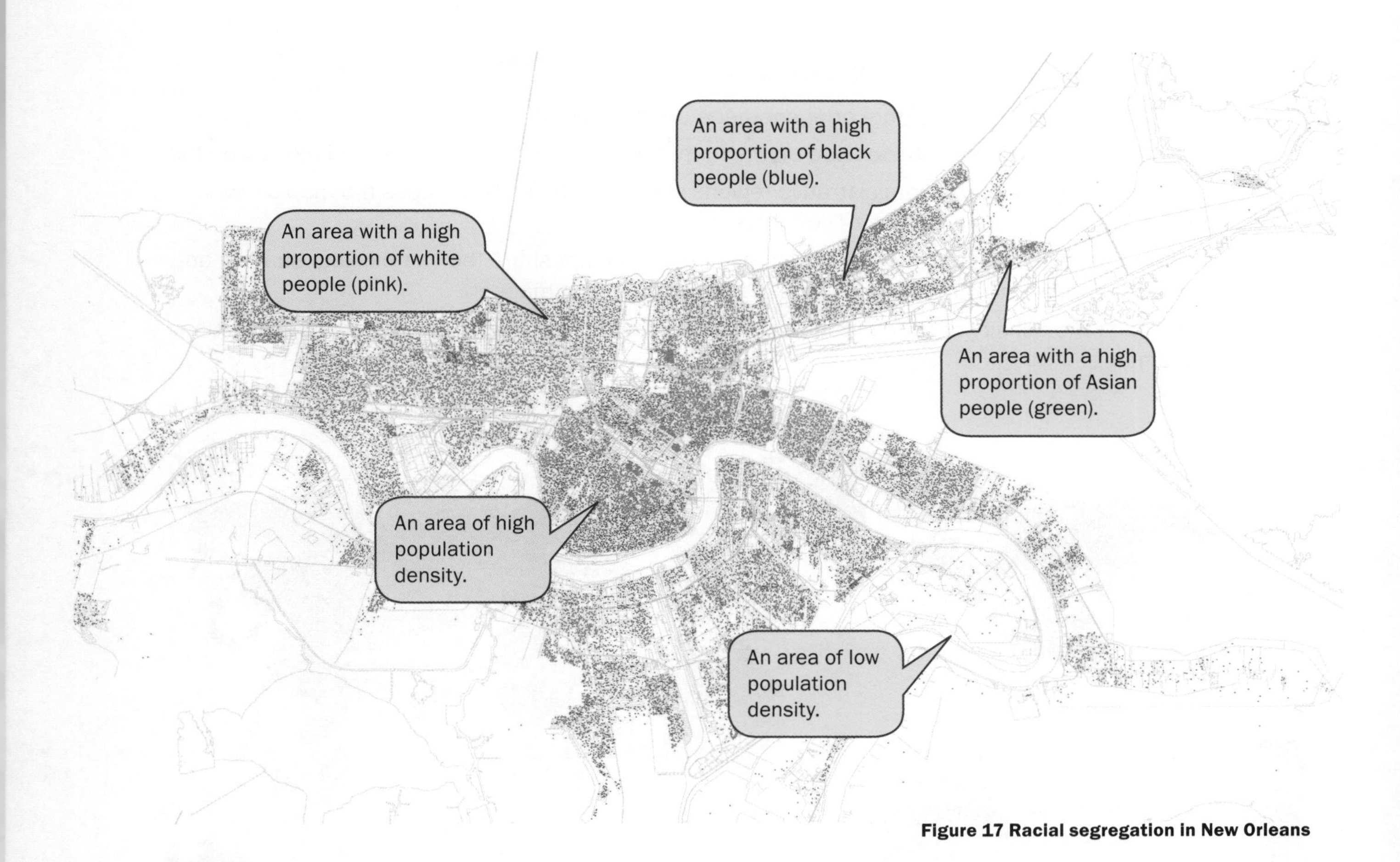

Figure 17 Racial segregation in New Orleans

To construct a dot map, follow the steps below:

1. Obtain a base map of the area or administrative regions you wish to show.
2. Study the data to be mapped and decide on a dot value. A rounded number should be chosen for the dot value to aid map interpretation.
3. Determine the number of dots required for each administrative region.
4. Decide on an appropriate dot shape and size. As a general guide, dots that are too small produce an overly sparse dot pattern, which is unnecessarily precise, while dots that are too large produce excessively dense dot patterns, which can mask the distribution of the mapped phenomenon. As a single dot will represent a set value, for clarity of presentation it is important that the dot size remains consistent throughout the map.
5. Place the correct number of dots within each administrative boundary as determined in Step 2.

Choropleth maps

A choropleth map is a thematic map that uses proportional shading to reveal spatial patterns within geographic data. They are often used in geography to:

- represent values or quantities per unit area of land such as local authority boundaries or administrative regions
- compare relative densities of areas
- show change over time by comparing maps from different areas.

However, choropleth maps are most frequently used to show variations in population characteristics across a region or continent.

ISBN: 9780170425285

To interpret a choropleth map, follow the steps below:

1. Identify the geographic feature or phenomena being mapped.
2. Verify the value of each shade used on the map. Reading the map's legend can help you do this.
3. Identify the scale of the administrative regions shown on the map (i.e. does the map show suburbs, census areas, states or countries?).
4. Using the key as a guide, identify the areas of the map that share the same colour shading and therefore the same quantity or volume of the feature being mapped.
5. Describe the density or concentration of the feature both within and between different areas of the map.

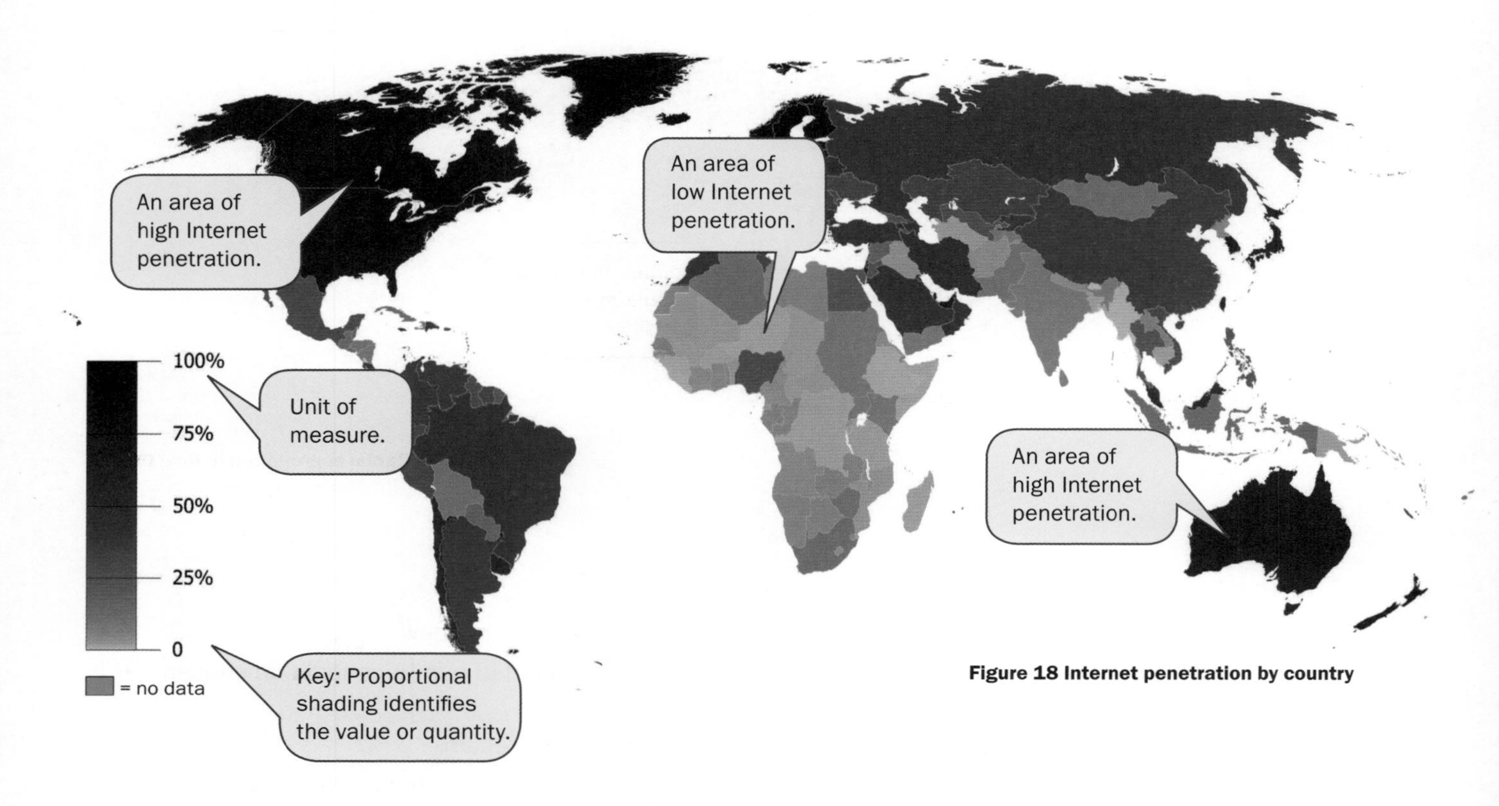

Figure 18 Internet penetration by country

To construct a choropleth map, follow the steps below:

1. Obtain a base map of the area or administrative regions you wish to show.
2. Examine the data to be presented and determine the range of data. This can be calculated by subtracting the lowest value in the data from the highest.
3. Divide the range by the number of categories you plan to use. Although most choropleth maps have four or five categories, more may be required if the data range is high. Regardless of the number of categories employed, ensure the value categories contain an even distribution of data (e.g. 1–100, 101–200, 201–300, 301–400, 401–500).
4. Assign a shade to each category. Typically, the darkest shade will be assigned to the highest-value category while the lightest shade is assigned to the lowest-value category.
5. Complete a category key on the map and sort the data into categories.
6. Shade in the administrative regions on your map according to your key.

ISBN: 9780170425285

Flow-line maps

Flow-line maps are designed to show the flow or movement of geographic phenomena from one location to another, such as the number of people in a migration, the volume of goods traded between regions, or water flows in a river basin.

When drawn well, flow-line maps allow the user to visualise the differences in magnitude or quantity of a range of flows. It achieves this by utilising arrows or lines of varying widths to represent the volume of objects being transferred between the place of origin and the place of destination. To read a flow-line graph correctly, the user must understand how to interpret the map's flow-line scale, which determines the value of the map's flow lines.

To interpret a flow-line map, follow the steps below:

1. Identify the geographic feature or phenomena being mapped.
2. Verify the value of each line or arrow used on the map. Reading the map's legend can help you do this.
3. Identify the scale of the administrative regions shown on the map (i.e. does the map show neighbourhoods, census areas, states or countries?).
4. Describe the direction and magnitude of the various movements of the geographic phenomena between different areas of the map.

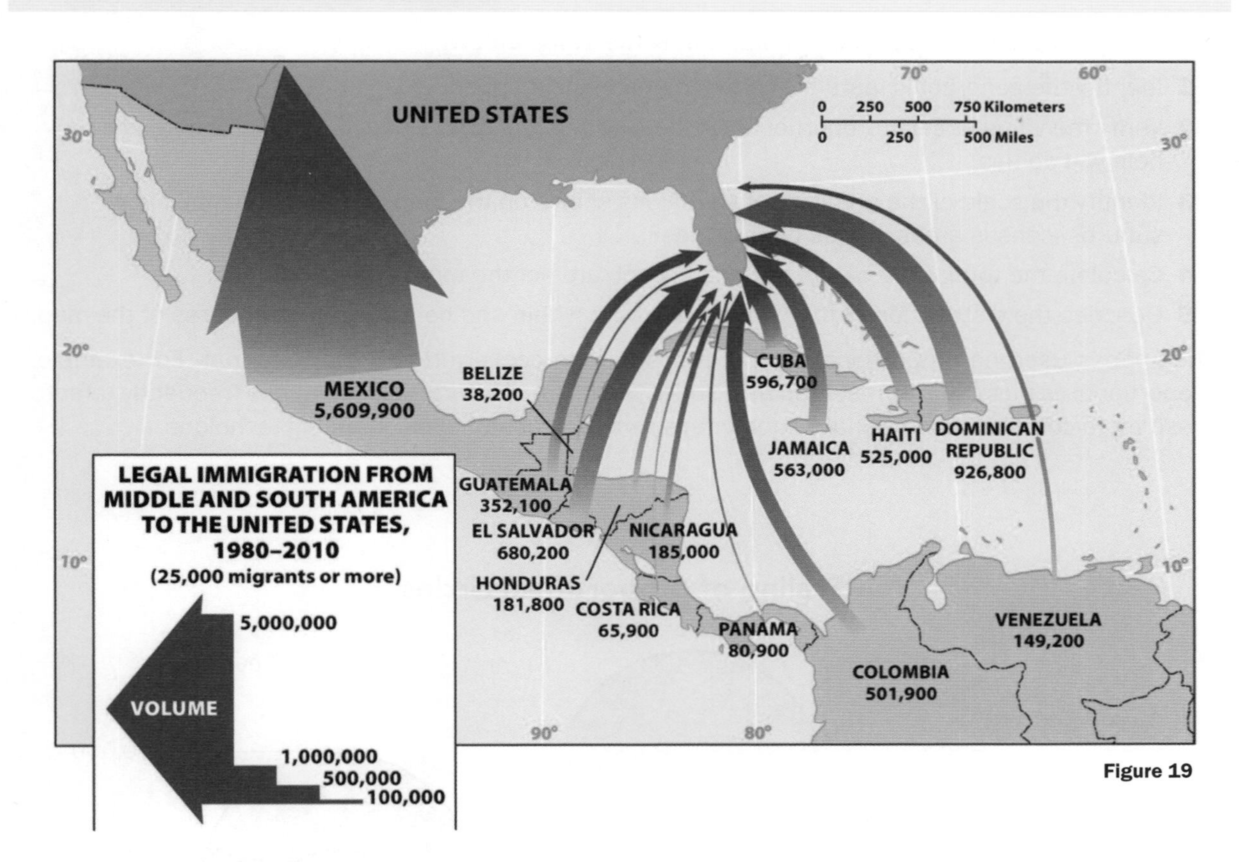

Figure 19

To construct a simple flow-line map, follow the steps below:

1. Obtain a base map of the region you wish to show.
2. Study the data to determine how thick each flow line will need to be to communicate the magnitude of each flow accurately. For example, 1 mm of line thickness could represent 100 data units.
3. Plot your data by drawing lines or arrows on the base map according to your predetermined scale. Position the tail of each flow line at the place of origin and add to it an arrowhead pointing to its destination.
4. Construct a key or legend to show the line scale, and the meaning of any symbols or shading used.

ISBN: 9780170425285

Proportional symbol maps

A proportional symbol map is a widely used thematic map that utilises symbols of different sizes to represent data associated with different locations on a map. Proportional symbol maps are frequently used in Geography because they are easy to read, allowing for the simple but effective identification of spatial patterns.

As with all thematic maps, employing proportional symbol maps has its advantages and a disadvantage.

Advantages:

- They are easy to interpret.
- Different symbols can be used to illustrate the distribution of multiple phenomena on the same map.
- They have the ability to show the different attributes of individual phenomena.

Disadvantage:

- Their interpretation is vulnerable to inaccurate perception of symbol size.

Almost any shape can be used in the construction of proportional symbol maps, including circles, squares, triangles and bars. Regardless of the shape employed, the size and area of the symbol must be in proportion to the data value it is representing.

To interpret a proportional symbol map, follow the steps below:

1. Identify the geographic feature or phenomena being mapped.
2. Verify the value of each proportional symbol used on the map. Reading the map's legend can help you do this.
3. Identify the scale of the administrative regions shown on the map (i.e. does the map show suburbs, census areas, states or countries?).
4. Calculate the total value of phenomena in each area of the map.
5. Describe the distribution of the phenomena both within and between different areas of the map.

The shapes used on proportional symbol maps can be converted to a graph format. For example, proportional circles can be presented as a series of pie graphs. NCEA examinations frequently refer to these more complex proportional symbol maps as a type of statistical mapping technique.

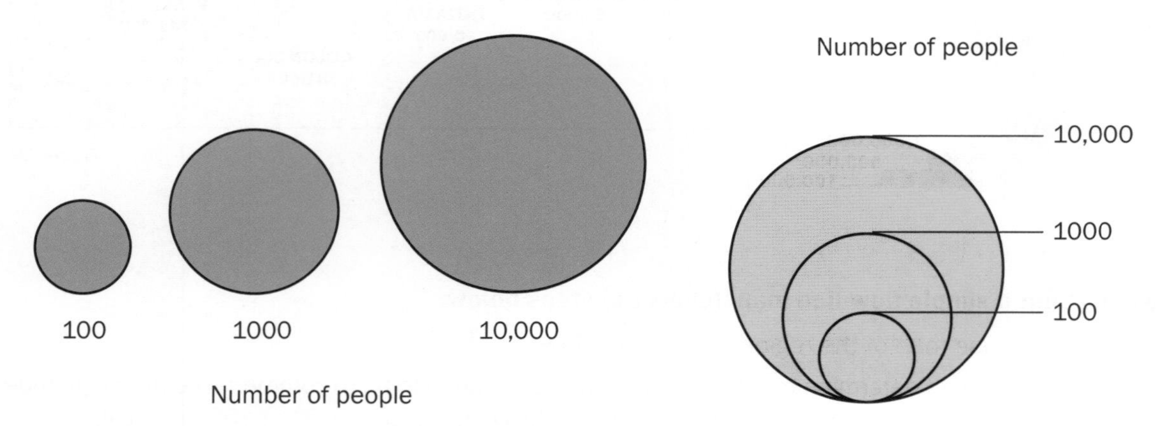

Figure 18

ISBN: 9780170425285

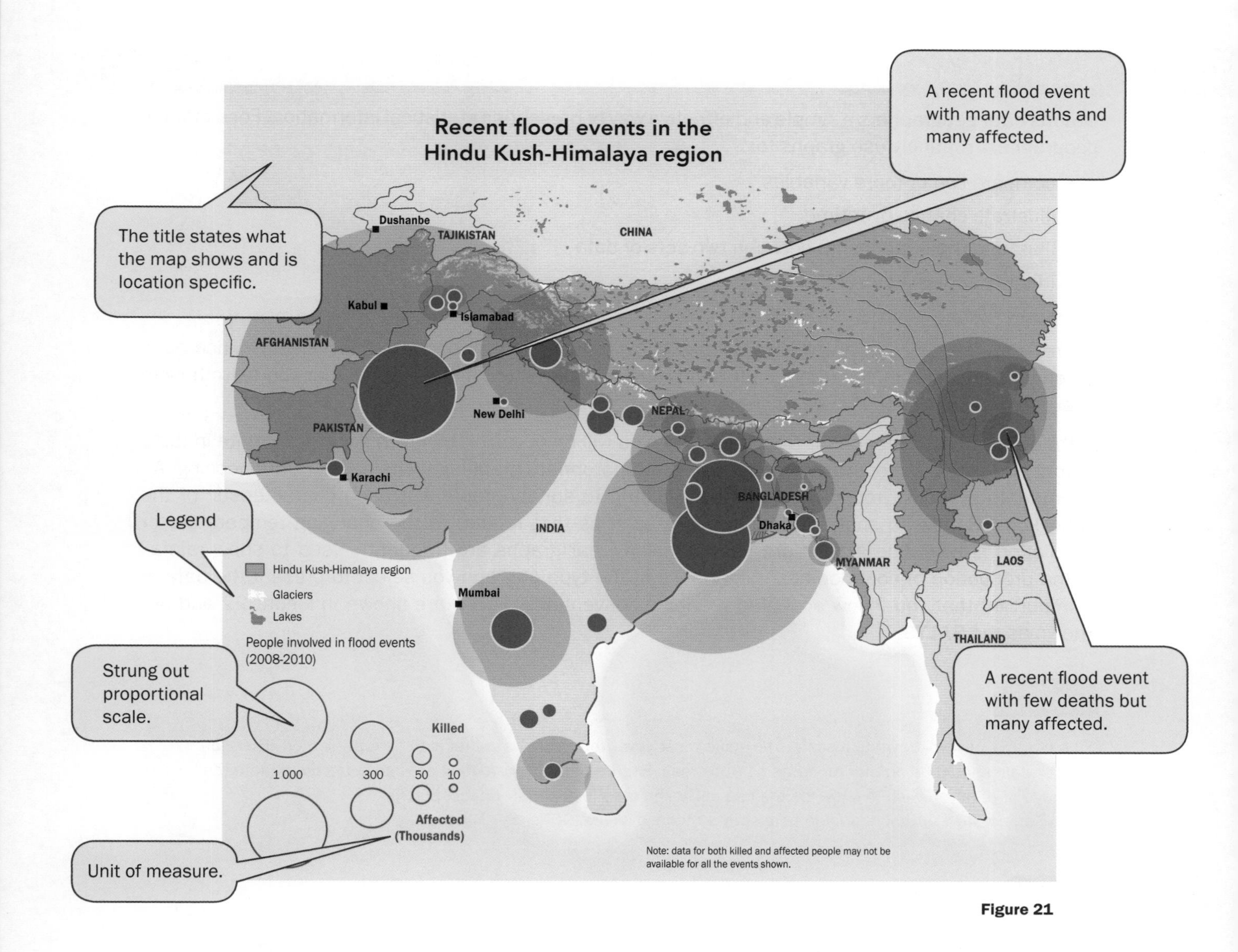

Figure 21

To construct a proportional symbol map, follow the steps below:

1. Obtain a base map of the area or administrative regions you wish to show.
2. Decide on a shape to use to represent the data you are planning to plot. Circles, squares, triangles and bars are commonly used in proportional maps, as it is not too difficult to change their area to represent different values.
3. To plot the data correctly, you will need to scale each symbol. Proportional circles are the most commonly used symbol, as they are easy to draw and can be scaled to represent values by changing the radius. For instance, the circle symbol representing a city of 100,000 would be ten times larger than the symbol for representing a town of 10,000.
4. Plot the data by drawing proportional symbols on the base map according to your predetermined scale. Take care to ensure the symbols do not overlap.
5. Construct a legend to show the proportional scale and meaning of any symbols or shading used.

ISBN: 9780170425285

Graphing skills

Graphs offer geographers a simple and effective way of presenting statistical information. For example, geographers routinely use graphs to:

- compare two or more variables
- illustrate change over time
- illustrate the relationship between two sets of data
- identify patterns or trends
- show how something is made up.

There are many types of graph, each designed to present information in a specific way. Since each graph type is suited to illustrating different types of information, it is essential that you learn how to interpret and construct a range of graph types.

In the external examination for Geography 3.4, you will on occasion be given geographic data in table format and asked to construct a graph using an appropriate graphing technique of your choosing. An appropriate graphical technique must be selected to suitably represent each characteristic of the data. For example, to show continuous data such as that related to the changing temperature or population growth, line graphs are used. Similarly, bar graphs are frequently used to show rainfall or the production of commodities. Regardless of the type of graph you select to present the data, it is important that you follow accepted graphing conventions. These are shown in Figure 22 and are known as the **SALTS**.

Scale

- A graph is always drawn to a scale. The scale must cover the entire data that is to be represented. Therefore, the scale should be neither too large nor too small. Ensure both axes are long enough to accommodate the range of data you wish to show, as the use of broken axes is generally unacceptable.
- The unit of measurement should be clearly stated e.g. years, 000s, %, millions, etc.

Axis

- Each axis must be clearly labelled to indicate what variable is being graphed. Depending on the type of data being graphed, the label can take the form of either categories or numerical amounts.

Legend

- The legend (or key) must clearly explain the colours, shades and symbols used in the graph. It is usually positioned at the lower left or lower right side of the graph.

Title

- The title defines the graph's location and purpose. It must be clear and include:
  - the name of the area
  - reference year of the data used.
- Example: New Zealand Inter-census Population Growth (1951-2018)

Source

- State where the data came from.

Figure 22 Graphing conventions

ISBN: 9780170425285

PQE can also be used to interpret the data and describe the trends and relationships found in graphs.

- **Pattern**: To describe the various patterns on a graph, look for a trend or correlation in the data, or concentrations of a particular variable.
- **Quantify**: Quantifying requires you to extract specific and accurate detail from the graph to support the patterns you have identified.
- **Exceptions**: In this step, account for elements or features in the data that do not fit the overall pattern or trend.

Bar graphs

Bar graphs are the simplest way to compare two sets of information. Generally, bar graphs consist of horizontal bars while column graphs use vertical bars. However, for the purposes of NCEA Geography you will not usually be required to make a distinction between horizontal bar and vertical column graphs.

To construct a simple bar graph, follow the steps below:

1. Decide what information is to be plotted on each axis. In most cases, you will plot the non-quantifiable variable (i.e. the one that does not change) on the *x*-axis (e.g. country names, age groups, or periods such as months or years) while the quantifiable data is normally plotted on the *y*-axis. It is for this reason the *y*-axis is sometimes referred to as the variable axis.
2. Bar graphs abide by the graphing convention that requires the *y*-axis (variable axis) to follow a constant number scale starting from zero (e.g. 0, 5, 10, 15 or 0, 10, 20, 30). You will therefore need to determine an appropriate scale for the variable axis.
3. Having determined the range and scale of the data to be plotted, use a ruler to construct the axes. Like most graphs, bar graphs have two axes: the *x*-axis is usually horizontal (i.e. runs across the bottom of the graph), while the *y*-axis is usually vertical (i.e. runs up the left-hand side of the graph). Ensure both axes are long enough to accommodate the range of data you wish to show.
4. Label each axis (including the units of measurement) and give the graph a title that clearly states what the graph illustrates. If appropriate, the title should also include location- and date-specific information.
5. Use a ruler to construct each bar. Ensure all bars are drawn with constant spacing and equal width.
6. Shade in each bar with a coloured pencil. If appropriate, label each bar or include a key if you are constructing a multiple bar graph (Figure 23).

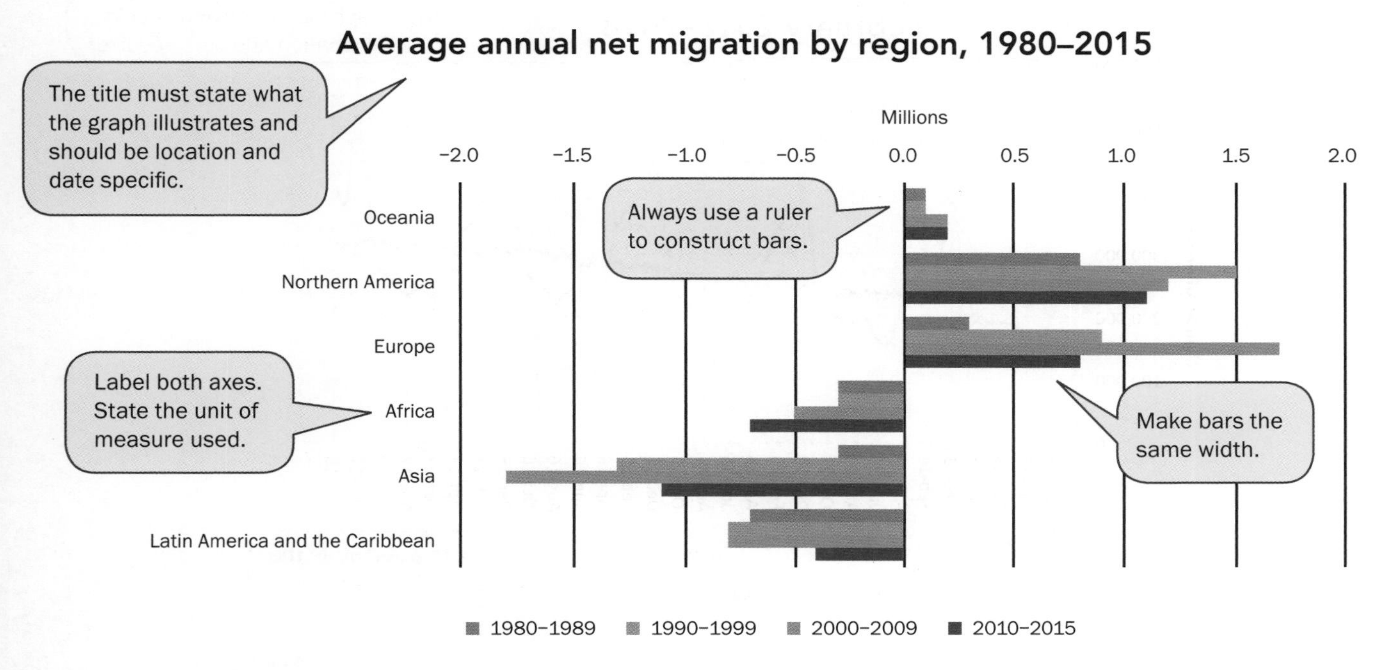

Resource 23 Multiple bar graph

Line graphs

Line graphs are especially useful in Geography as they are easy to create, and their visual characteristics reveal data trends clearly.

Like bar graphs, line graphs provide a visual representation of how two variables (shown on the *x*-axis and *y*-axis) relate. The vertical *y*-axis in a line graph usually indicates quantity (e.g. population size or change, volume) or percentage in the case of a compound graph, while the horizontal *x*-axis often measures units of time (e.g. years). As a result, the line graph is often used to show change over time. For example, if you wanted to graph changes in the crude birth rate over time, you could measure the time variable in years along the *x*-axis, and birth rate (per 1000) along the *y*-axis.

Projected data is usually shown on a line graph by a dotted or dashed line.

To construct a simple line graph, follow the steps below:

1. Decide what information is to be plotted on each axis. In most cases, you will plot the non-quantifiable variable (i.e. the one that does not change) on the *x*-axis (e.g. country names, age groups, or periods such as months or years) while the quantifiable data is normally plotted on the *y*-axis. It is for this reason the *y*-axis is sometimes referred to as the variable axis.
2. Like bar graphs, line graphs also abide by the graphing convention that requires the axes to follow a constant number scale starting from zero (e.g. 0, 5, 10, 15 or 0, 10, 20, 30). You will therefore need to determine an appropriate scale for the variable axis.
3. Having determined the range and scale of the data to be plotted, use a ruler to construct the axes. Like most graphs, simple line graphs have two axes: the *x*-axis is usually horizontal (i.e. runs across the bottom of the graph), while the *y*-axis is usually vertical (i.e. runs up the left-hand side of the graph). Ensure both axes are long enough to accommodate the range of data you wish to show.
4. Label each axis (including the units of measurement) and give the graph a title that clearly states what the graph illustrates. If appropriate, the title should also include location and date specific information.
5. Next, plot each value on the graph, and then join the points together either with a ruler for a straight-line curve or freehand if a smooth curve is required.
6. If you are constructing a multiple line graph, it is recommended that you also include a key.

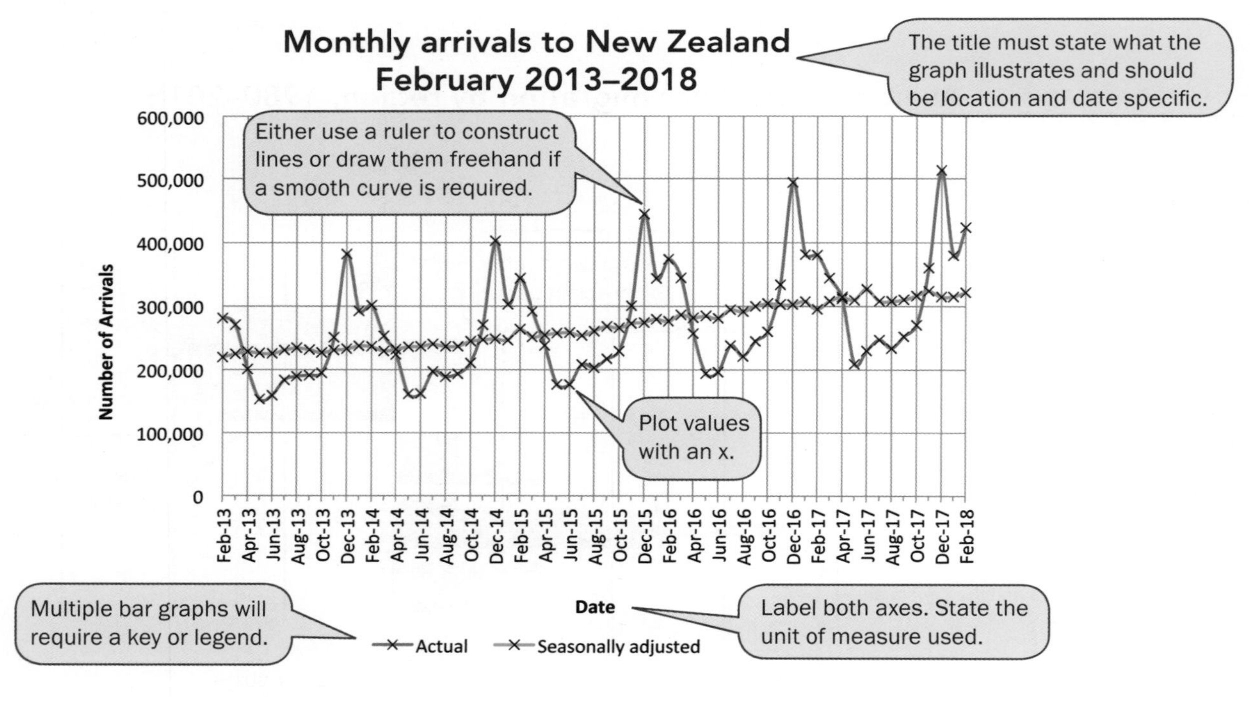

Figure 24 Line graph

ISBN: 9780170425285

Proportional graphs

A proportional graph is a useful way of summarising data that has been categorised (e.g. religious affiliation) or represents different values of a given variable (e.g. the age distribution of a population). Pie graphs and percentage bar graphs are the most common ways of presenting proportional data.

Pie graphs

To construct a pie graph, follow the steps below:

1. Use a compass or stencil to construct a circle. Then draw a line from the centre of the circle to the 12 o'clock position.
2. Convert each percentage value into degrees by multiplying each variable by a factor of 3.6. For example, in Figure 25 the exotic grasslands portion of the graph on the left-hand side was calculated by multiplying the percentage of land covered by exotic grasslands by 3.6 to represent 140.4 degrees (i.e. 39% x 3.6 = 140.4°).
3. Working clockwise from the 12 o'clock position, use a protractor to plot each segment of the pie beginning with the largest category followed by the second largest category and so on. If applicable, plot the 'Other' category last.
4. Shade in each segment with coloured pencils. Label each segment or include a key.
5. Give the graph a title that clearly states what the graph illustrates. The title should also include any location and date specific information.

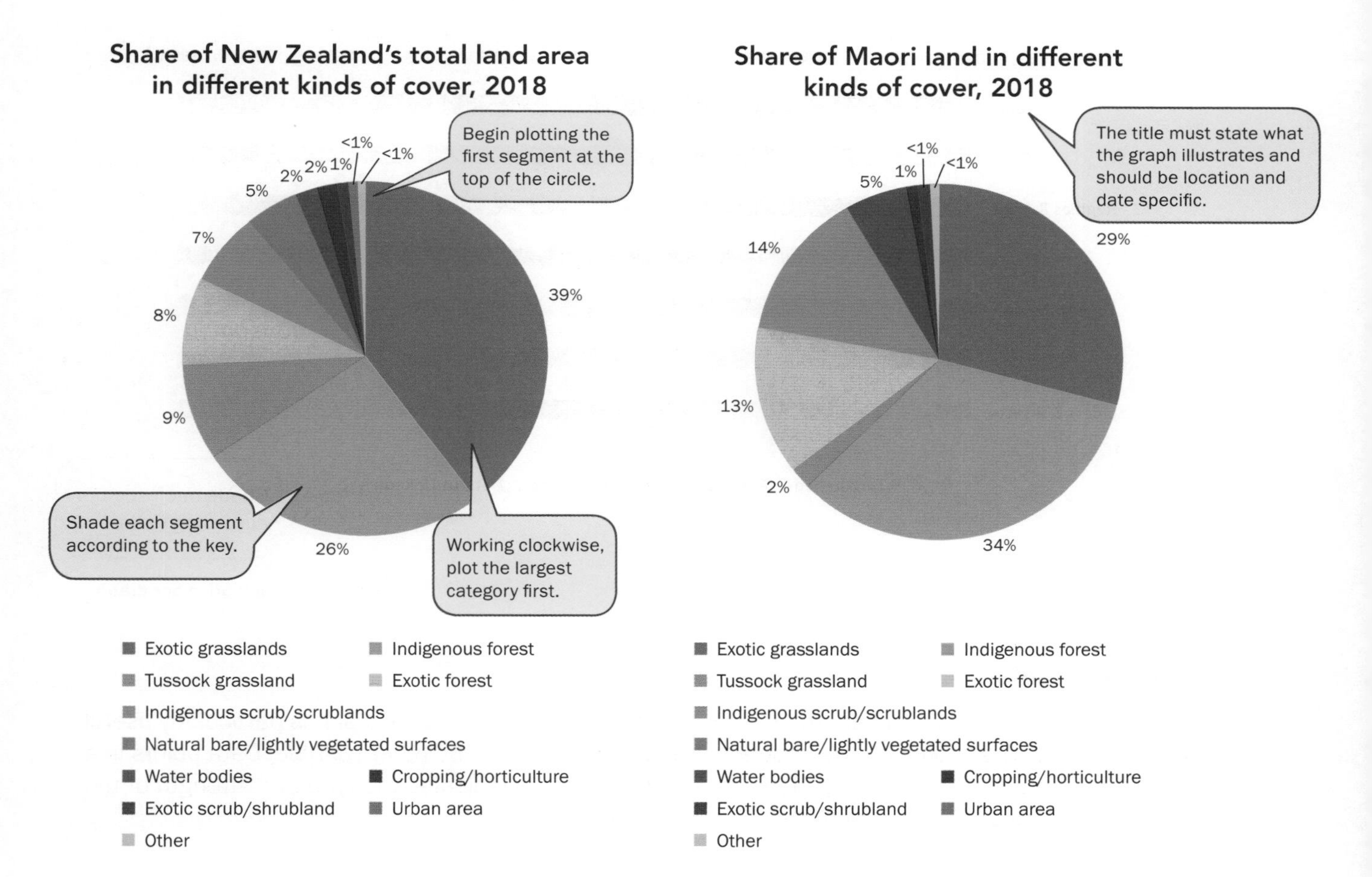

Figure 23 Pie graphs

ISBN: 9780170425285

Cumulative or percentage bar graphs

Cumulative bars are usually drawn 100 mm long to allow for easy conversion from percentage to segment length (Figure 24).

To construct a cumulative bar graph, follow the steps below:

1. Use a ruler to construct a horizontal bar measuring 100 mm by 10 mm.
2. Convert each percentage variable into an equivalent proportional distance. For example, a variable of 25% would be equal to a segment length of 25 mm, and a variable of 14% would be equal to a segment length of 14 mm.
3. Working from the left-hand end of the bar, measure and plot each segment beginning with the largest category followed by the second largest category and so on. If applicable, plot the 'Other' category last.
4. Shade in each segment with coloured pencils. Label each segment or include a key.
5. Give the graph a title that clearly states what the graph illustrates. If appropriate, the title should also include location and date specific information.

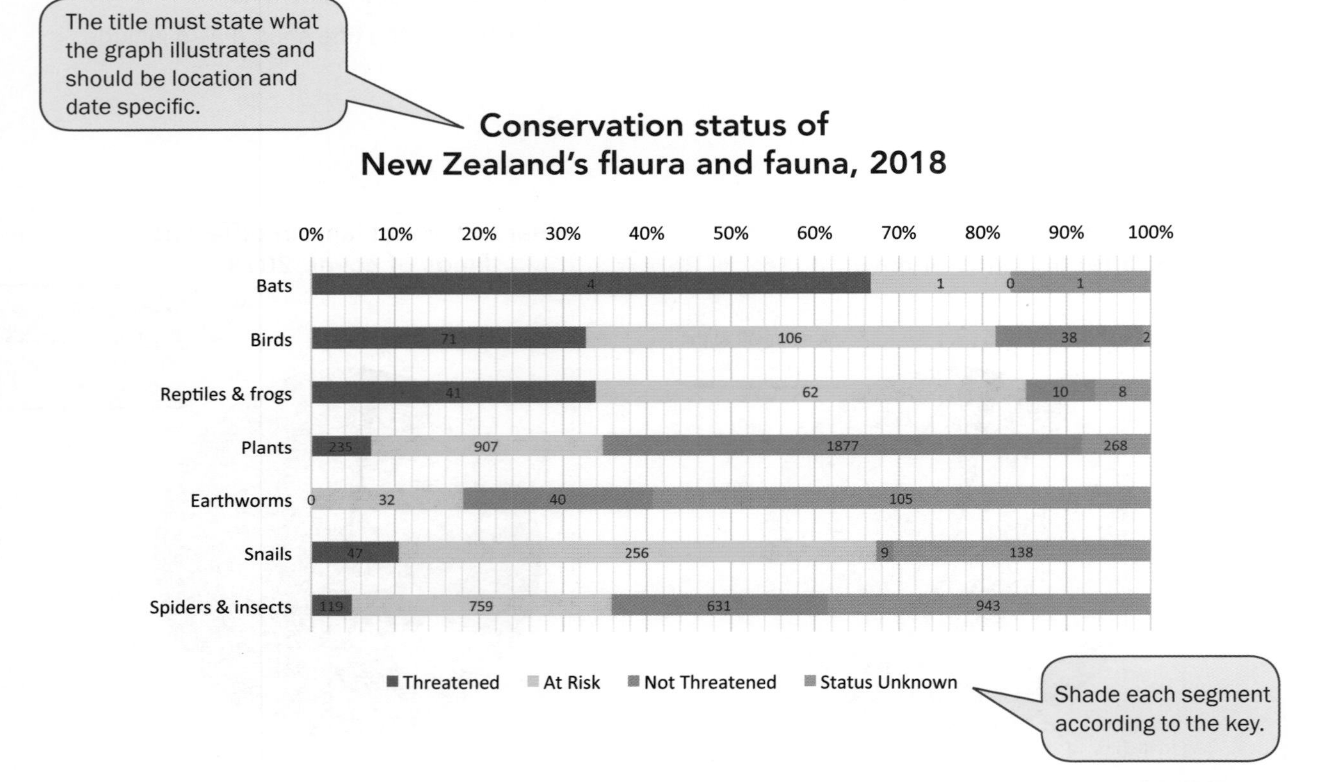

Figure 24 Cumulative bar graph

Scatter graphs

A scatter graph is used to present the measurements of two related variables. It is particularly useful when one variable is thought to be dependent upon the values of the other variable. Data points in a scatter graph are plotted but not joined. The resulting pattern indicates the type and strength of the relationship between the two variables (Figures 25 and 26).

ISBN: 9780170425285

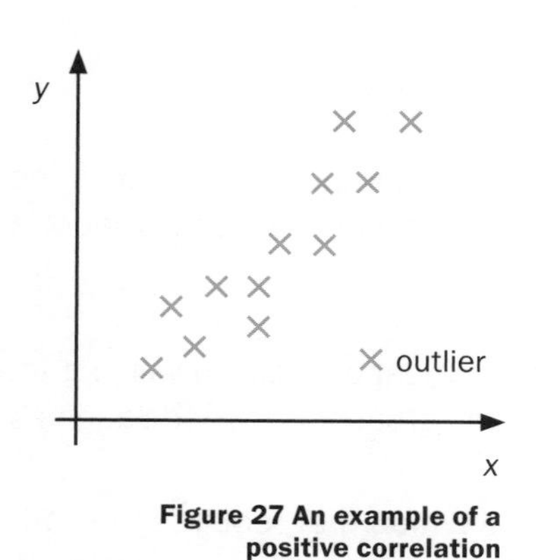

Figure 27 An example of a positive correlation

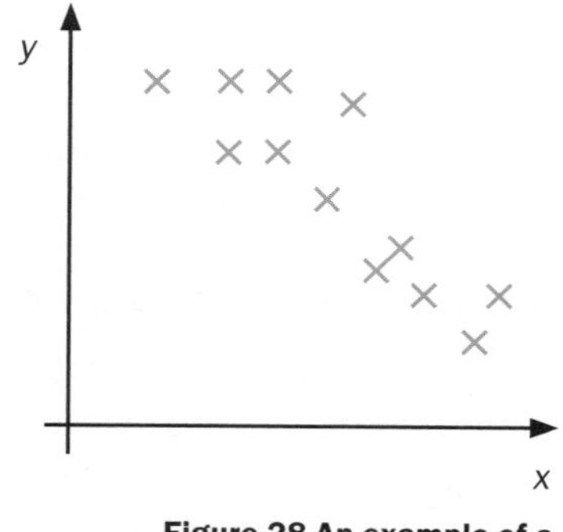

Figure 28 An example of a negative correlation

To construct a scatter graph, follow the steps below:

1. Decide what information is to be plotted on each axis. In most cases, you will plot quantifiable data on both the *x*-axis and the *y*-axis.
2. Like line and bar graphs, scatter graphs abide by the graphing convention that requires the axes to follow a constant number scale starting from zero (e.g. 0, 5, 10, 15 or 0, 10, 20, 30). You will therefore need to determine an appropriate scale for the two axes. However, if the range of data is too broad, one axis may employ a logarithmic scale.
3. Having determined the range and scale of the data to be plotted, use a ruler to construct the axes. Like most graphs, simple line graphs have two axes: the *x*-axis is usually horizontal (i.e. runs across the bottom of the graph), while the *y*-axis is usually vertical (i.e. runs up the left-hand side of the graph). Ensure each axis is long enough to accommodate the range of data you wish to show.
4. Label each axis (including the units of measurement) and give the graph a title that clearly states what the graph illustrates. If appropriate, the title should also include location and date specific information.
5. Next, plot the intersection of each value on the graph with an x. It is important that you do not join the points together.
6. Draw a line of best fit.

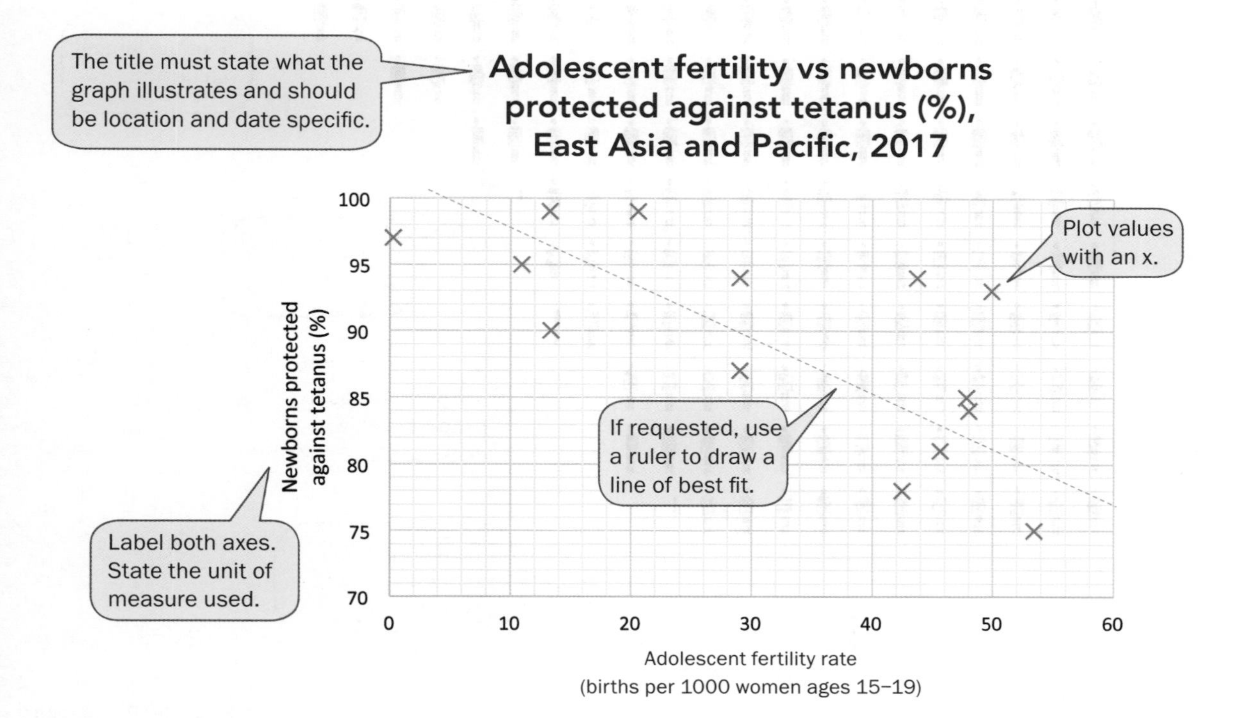

Figure 29 Scatter graph

Pictographs

A pictograph is a pictorial representation of statistical information where each value is represented by a proportional number of pictures or symbols. A pictograph has a form similar to that of a bar graph.

To construct a pictograph, follow the steps below:

1. Decide what information is to be plotted on each axis. In most cases, you will plot the non-quantifiable variable on one axis (e.g. country names, age groups, or periods such as months or years) and quantifiable data on the other.
2. Like bar graphs, pictographs abide by the graphing convention that requires the variable axis to follow a constant number scale starting from zero (e.g. 0, 5, 10, 15 or 0, 10, 20, 30). You will therefore need to determine an appropriate scale for the variable axis.
3. Having determined the range and scale of the data to be plotted, use a ruler to construct the axes. Like most graphs, pictographs have two axes: the *x*-axis is usually horizontal (i.e. runs across the bottom of the graph), while the *y*-axis is usually vertical (i.e. runs up the left-hand side of the graph). Ensure each axis is long enough to accommodate the range of data you wish to show.
4. Label each axis (including the units of measurement) and give the graph a title that clearly states what the graph illustrates. If appropriate, the title should also include location and date specific information.
5. Choose a symbol to represent the variable you want to plot. Then decide the quantity that the symbol will represent. For example, in Figure 30 the symbol represents one million people. When plotting the symbols onto your graph, ensure they are drawn with constant spacing and width.
6. If appropriate, label each bar and include a key.

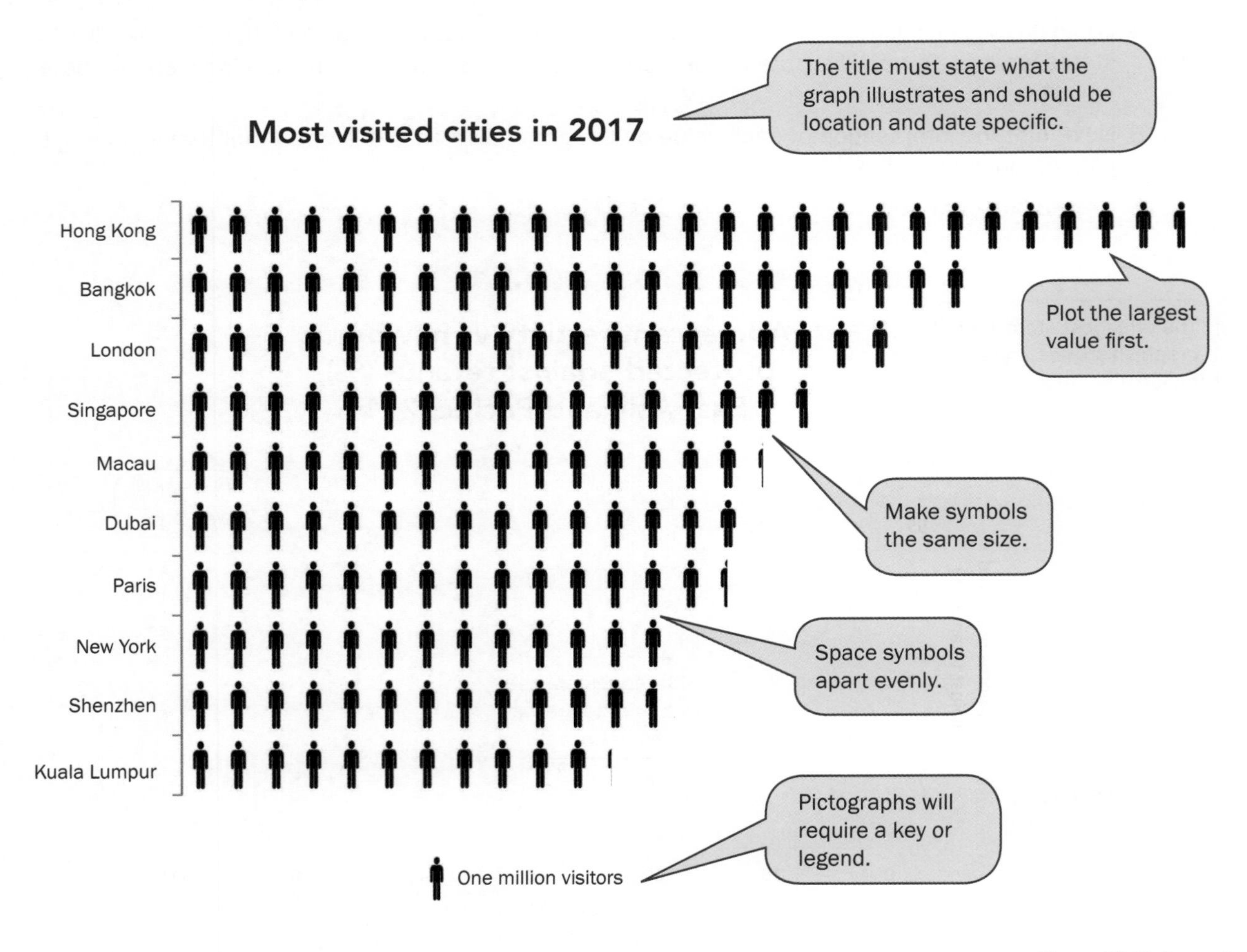

Figure 30 Pictograph

ISBN: 9780170425285

Climate graphs

A climate graph shows average temperature and rainfall experienced at a particular location throughout the year. It consists of a bar graph showing average monthly rainfall and a simple line graph showing average monthly temperature.

To construct a climate graph, follow the steps below:

1. A climate graph has one horizontal axis bounded by two vertical axes. To construct the horizontal axis, divide the axis into 12 even segments to represent the months of the year.
2. Place the rainfall scale (mm) on the left-hand side of the graph and the temperature scale (°C) on the right-hand side of the graph. Determine an appropriate scale for the two vertical axes and follow a constant number scale starting from zero for each axis (e.g. 0, 5, 10, 15 or 0, 100, 200, 300).
3. Label each axis (including the units of measurement) and give the graph a title that clearly states what the graph illustrates. The title should include location-specific information and may even include the location's latitude and longitude.
4. Plot rainfall data for each month and colour each plotted bar blue.
5. Plot the average temperature data for each month with an x. Ensure that each temperature value is positioned at the centre of each month. Join the data points with a smooth red curve.

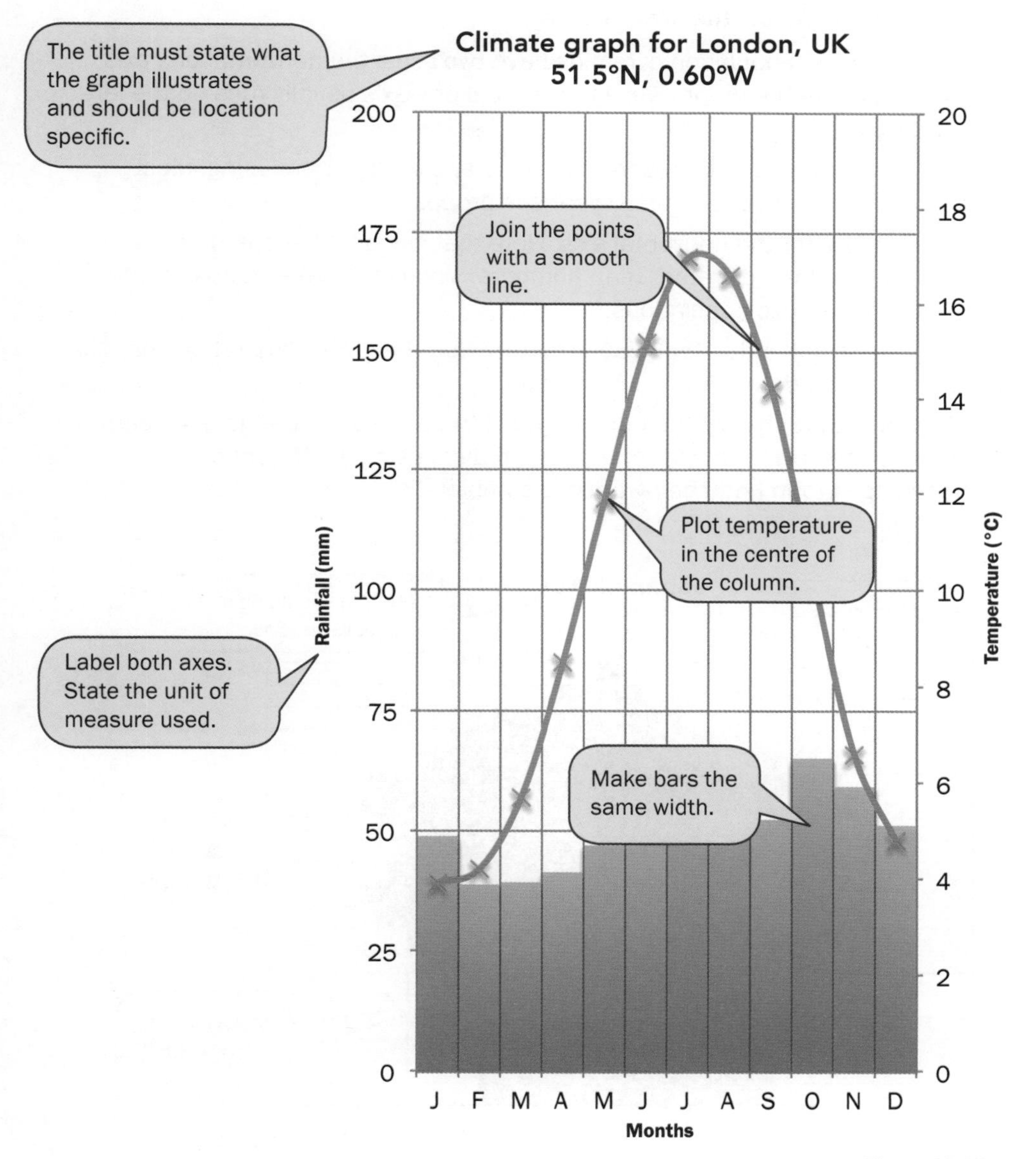

Figure 31 Climate graph

ISBN: 9780170425285

Population pyramids

A population pyramid is a special type of horizontal bar graph that shows the age-sex structure of a population in one-year, five-year, or 10-year age groups.

The shape of a population pyramid reflects the influence of births, deaths and migration on a population over time and shows whether a population is expanding, stable or likely to decline (Figure 32). In general, a population with a high birth rate and low death rate has a broad-based, triangular-shaped pyramid. Populations with low birth rates and low death rates are usually narrower at the base and have straighter sides.

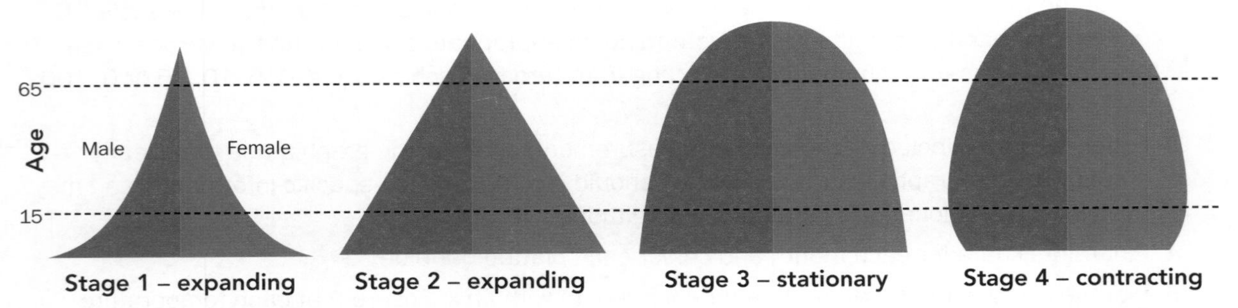

Figure 32 Population pyramid shapes

To construct a population pyramid, follow the steps below:

1. Use a ruler to construct the axes. Population pyramids have two axes: the horizontal or *x*-axis usually runs across the bottom of the graph, while the vertical or *y*-axis usually runs up the left-hand side or centre of the graph.
2. Divide the vertical axis into segments to correspond with the age data you are using. Most population pyramids use one-year, five-year, or 10-year age groups.
3. Determine an appropriate scale for the horizontal axis. Note that it is best to construct population pyramids using percentages rather than numbers, since this makes it possible to compare countries with different-size populations.
4. Label each axis and give the graph a title that clearly states what the graph illustrates. The title should also include location- and date-specific information.
5. Beginning at the bottom of the graph, plot the percentage of the population that is 0–4 years and male. Shade this bar on the pyramid and repeat for females, using a different colour. Repeat this step for each age group until the pyramid is complete.

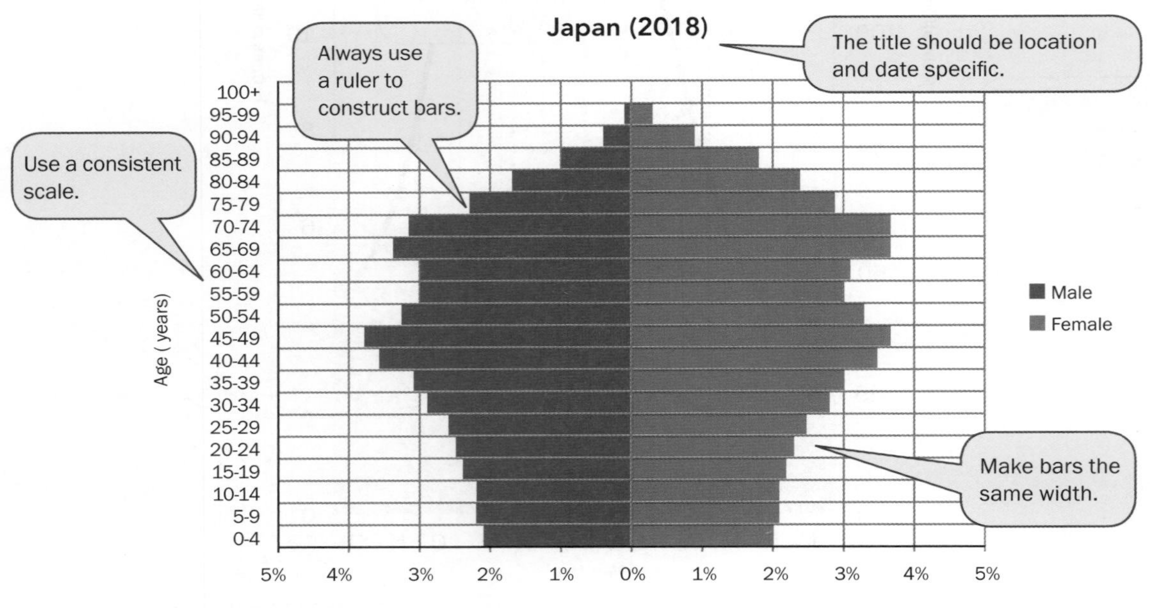

Figure 33

ISBN: 9780170425285

Visual interpretation skills

Visual images can be used to illustrate almost any aspect of Geography. As a student of Geography, you may on occasion be asked to identify and interpret what you can see in an image, or asked to compare the information from an image with that from a map.

Photographic interpretation

Geographers regularly use photographic images to record information about a setting. Photographic images are a useful tool as they provide a visual record of natural and cultural features of the environment, and help us to understand the way different elements of the environment relate or interact.

Photographs also allow geographers to study how environments change over time. By comparing photographs taken at different times, it is possible to analyse change in any one environment.

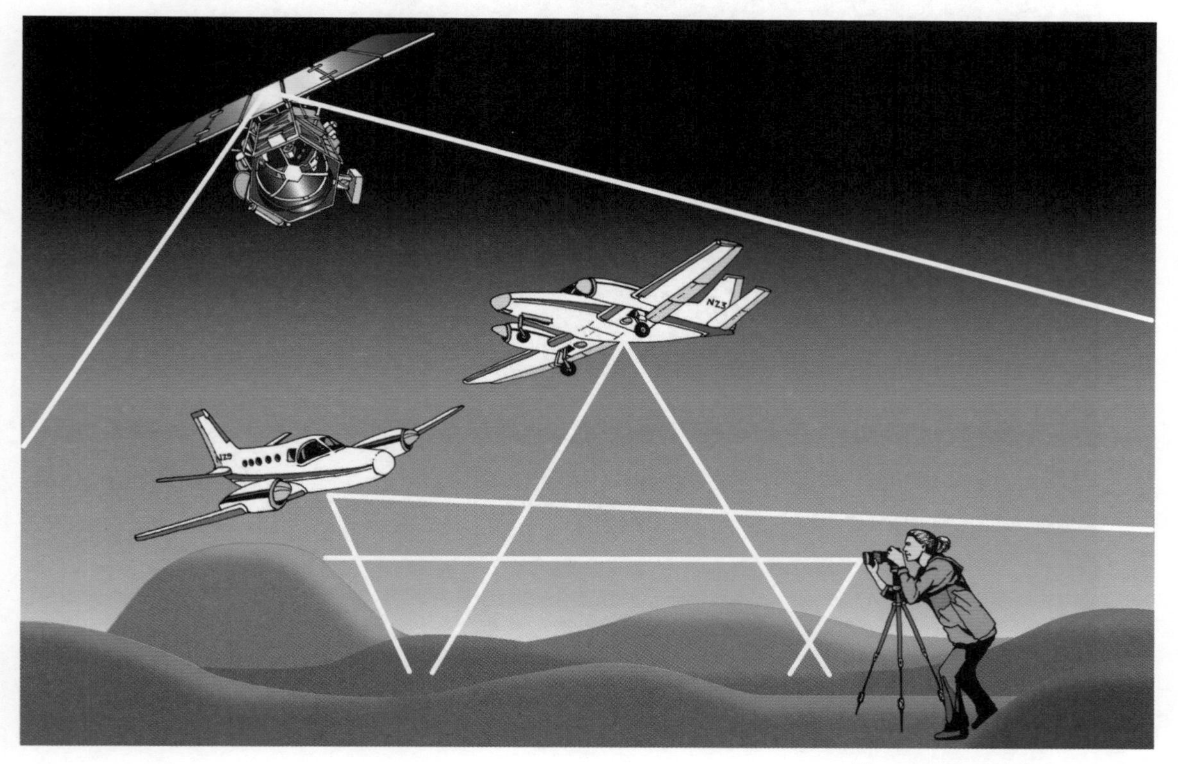

Figure 34 Types of photographs

Three types of photographs are useful to geographers:

- **Ground-level photographs** are images taken from the ground to maximise the horizontal view. With ground-level photographs, foreground features appear larger than background features.
- **Aerial photographs** are images taken of the earth's surface from the air. They show a bird's-eye view either from directly above (vertical) or from an oblique angle. Oblique photographs have an advantage over vertical photographs in that they show both the tops and sides of objects, making them easier to identify. The main disadvantage of oblique photographs is that they do not have a consistent scale.
- **Satellite images** are created from data collected from satellites orbiting the earth. They usually use artificial colours. With satellite images, spatial patterns are clearly visible over a large area (Figure 35).

ISBN: 9780170425285

To interpret a photograph, follow the steps below:

1. Determine whether the photograph shows an aerial or ground-level perspective. If the perspective is aerial, determine if its orientation is vertical or oblique.
2. Identify any visual clues that help identify the photograph's setting and location.
3. Identify the main natural (e.g. relief and drainage features, vegetation, soil and climate) and cultural (e.g. patterns of settlement, transportation networks and other land uses) features of the environment shown in the photograph.
4. Examine the way different features shown in the photograph interact. For example, the photograph might show settlement along a coastline or on the floodplain of an adjacent river.

Figure 35 Satelite image of Dubai, United Arab Emirates

Geographical information systems

A geographical information system (GIS) is a computer system that is capable of collecting, storing, analysing and displaying geographical data. GISs help geographers better understand spatial associations and patterns.

GIS have many different applications outside Geography and are routinely used in smartphone applications, planning, communications, and transport and logistics. A car's navigation system uses GIS data to enable drivers to make smarter decisions as they navigate through the real world.

ISBN: 9780170425285

Numeracy skills

Numeracy refers to the skill of being able to understand and work with numbers. However, just because numeracy deals with numbers does not mean that its use is limited to the subject of mathematics. Geographers routinely use numbers to help them identify and interpret patterns and trends in geographic phenomena.

The ability to interpret and then analyse statistical information is an essential skill for a geographer to possess. In Level 3 NCEA Geography, you need to know how to work with numbers to calculate means, modes, ranges and percentage change.

Calculating mean, mode and range

The most basic of numeracy skills is the ability to understand and calculate statistical means, modes and ranges.

1 When working with a data set, the **mean** (or arithmetic average) is calculated by finding the sum of all the given values divided by the total number of values. It is expressed by formula as:

$$\overline{x} = \frac{\sum x}{n}$$

where $\overline{x}$ = mean, $\sum$ = the sum of, x = the value of an individual variable and n = the number of values in the data set. Put in another way, the formula can be presented as a statement:

mean = sum of all the observed values ÷ number of observations

For example, in a rural village of 10 households, the number of people living in each household is 7, 5, 0, 7, 8, 5, 5, 4, 5 and 2.

Mean = (7 + 5 + 0 + 7 + 8 + 5 + 5 + 4 + 5 + 2) ÷ 10
= 48 ÷ 10
= 4.8

2 The **mode** is the most frequently observed data value. There may be no mode if no value appears more than any other.

mode = the most frequently observed data value

In the previous example, the value of the mode is 5, as four households contain five inhabitants. This number occurs more frequently than any other data value.

3 The data **range** refers to the difference between the highest and the lowest observed data value.

range = the highest observed data value − the lowest observed data value

In our previous example, the range would be calculated as follows:

Range = 8 − 0
= 8

Calculating percentage change

When working with statistics, it is sometimes useful to be able to quantify the rate of proportional change between an original value and its new value. A popular mathematical technique for measuring change between two values is to calculate its percentage change.

To calculate percentage change, apply the following formula:

$$\textbf{percentage change} = \frac{\textbf{(difference between the two values)}}{\textbf{original value}} \times \frac{\textbf{100}}{\textbf{1}}$$

For example, in 1960 New Zealand's population was 2.37 million. In 2010, it was estimated at 4.38 million. The percentage increase between 1960 and 2010 would be calculated as follows:

$$\text{percentage change} = \frac{(4.38\text{ million} - 2.37\text{ million})}{2.37\text{ million}} \times \frac{100}{1} = 85\%$$

ISBN: 9780170425285

Examination ready

Instruction or command words used in Level 3 examination questions will also reflect the expectation of longer written explanations such as the need to *justify* or *evaluate* rather than simply *describe* or *explain*, as was often required at earlier levels. Recognising and understanding the requirements of the command word in a question is essential to answering it successfully (Figure 36).

Command word	Meaning	Example question
Analyse	Deconstruct (be breaking down) date (information) found in a map, graph, diagram, table or text to identify patterns, trends, connections and relationships.	Use specific information from the resources to analyse the pattern of internal migration in China.
Annotate	Add brief notes to a diagram or graph.	Draw an annotated diagram to show the key features of a coastal environment.
Compare	Give an account of the similarities between two or more items or situations, referring to all of them throughout.	Compare coastal landforms on Auckland's west and east coasts.
Compare and contrast	Give an account of the similarities and differences between two or more items or situations, referring to all of them throughout.	Compare and contrast the tourist facilities in two different cities.
Contrast	Give an account of the differences between two or more items or situations, referring to them throughout.	Contrast the effectiveness of anti-natalist and pro-natalist population policies.
Describe	Identify and give an account of a given situation, event, pattern, trend, process or feature.	Describe the trend in population growth shown on the graph.
Discuss	Give a considered and balanced review that includes a range of arguments or factors. Opinions or conclusions should be presented clearly and supported by examples.	'Floods are more hazardous than earthquakes.' Discuss this statement.
Draw	Represent using a labelled, accurate diagram or graph.	Draw a suitable graph to compare how per capita consumption of renewable energy has changed in the countries of the OECD since 2000.
Evaluate	Make an appraisal weighing up the strengths and weaknesses of the available options.	Evaluate the sustainability of tourism development in Queenstown.
Examine	Consider an argument or concept in a way that uncovers assumptions and interrelationships of an issue.	Examine why some countries want to reduce their dependence on fossil fuels.
Explain	Give a detailed account including reasons or causes.	Explain the relationship between fertility and female literacy.
Identify	Give an answer from a number of possibilities.	Identify the year in which the number of visitors to New Zealand exceeded the number of departures.
Justify	Give valid reasons or evidence/examples to support an answer or conclusion.	Justify the position taken by anti-globalisation movements.
Outline	Give a brief account or summary.	Outline two changes in literacy rates shown in the graphs.
State	Give a specific name, value or other brief answer – no need for explanation or calculation.	State the three components that are used to calculate the Human Development Index.
Suggest	Propose a solution, hypothesis or other possible answer.	Suggest possible reasons for the changes in oil consumption between 1970 and 2010 shown on the graph.
To what extent	Consider the strengths/merits of an argument or concept. Opinions and conclusions should be presented clearly and supported by examples and sound arguments.	To what extent has climate shaped the environment of the Atacama Desert?

Figure 36 Command terms

The chapters that follow offer you the opportunity to practise the types of questions you can expect in the final Geography 3.4 examination. Therefore, as will be the case in the final examination, you should take approximately 60 minutes to complete each chapter in this booklet. You are also encouraged to use coloured pencils (black, blue, green, red, brown, and yellow), a calculator, and a ruler where appropriate when answering the questions in this booklet. You should use coloured pencils when constructing diagrams and maps. However, labels and annotations on these diagrams and maps must be in pen. Also, note, that in the final examination, written work done in pencil will not be eligible for reconsideration.

ISBN: 9780170425285

From Recovery to Rebuild

On 22 February 2011, a 6.3 M_L (local magnitude) earthquake devastated the city of Christchurch, New Zealand's second-largest city. The impact of the earthquake on Christchurch and the wider Canterbury region was considerable; killing 185 people, injuring about 5800, damaging more than 100,000 homes, and inflicting $40 billion worth of damage. Christchurch's central business district (CBD) was worst affected with more than 70 percent of buildings located within it either destroyed or damaged beyond repair.

With the recovery phase all but over, the people of Christchurch have an unprecedented opportunity to rebuild their city in a way that integrates cutting-edge design with the latest ideas in urban planning. However, from similar experiences overseas, we know the transition from recovery to rebuild could take 20 years or more.

Q1 Natural environment of Christchurch, Canterbury

Learning Activities

Applying a geographic concept: Environments

Environments may be natural and/or cultural. They have particular characteristics and features, which can be the result of natural and/or cultural processes. The particular characteristics of an environment may be similar to and/or different from another.

Refer to the definition of environments above and **Resources 1A** and **1B** to answer the following questions.

a Describe in detail the location of Christchurch according to its geological setting.

ISBN: 9780170425285

Learning Activities

b Describe in detail one aspect of the natural environment of Christchurch.

c Suggest reasons why the people of Christchurch are vulnerable to the threat of earthquakes.

ISBN: 9780170425285

Q2 Population change

Applying a geographic concept: Change

Change involves any alteration to the natural or cultural environment. Change can be spatial and/ or temporal. Change is a normal process in both natural and cultural environments. It occurs at varying rates, at different times and in different places. Some changes are predictable, recurrent or cyclic, while others are unpredictable or erratic. Change can bring about further change.

Post-earthquake statistics reveal that Christchurch's population dropped by almost 10,600 people in the months immediately after the February 2011 earthquake, however, because of subsequent natural population growth, the overall drop for the year was only 8900. The 2.4 percent population decline was a marked about-turn for the city, which had an annual pre-quake population growth of 1 percent (about 3800 people).

a Refer to **Resource 1C** to answer this question.

i Use an appropriate statistical mapping technique to display the number of people who moved out of the Canterbury region to other regions of New Zealand after the February 2011 earthquake. Use appropriate mapping conventions.

ISBN: 9780170425285

Learning Activities

ii With reference to the statistical map you have drawn, describe in detail the effect the Christchurch earthquake has had on migration patterns. Tip: use PQE.

b Refer to the definition of change and **Resources 1C – 1F** to answer the following questions.

i Compare the impact of the 22 February 2011 earthquake on the population of Christchurch to that of other comparative hazard events.

ISBN: 9780170425285

Learning Activities

ii Evaluate how the demographic environment of Christchurch changed as a result of earthquake activity.

ISBN: 9780170425285

Q3 The process of recovery

Learning Activities

Applying a geographic concept: Processes

A process is sequence of actions, natural and/or cultural, that shape and change environments, places and societies. Some examples of geographic processes include erosion, migration, desertification and globalisation.

Christchurch's post-earthquake rebuild refers to the restoration and enhancement of the city's natural and cultural environments. It is inherently future-focused in that it does not seek to return Christchurch to how it was before the February 2011 earthquake but instead seeks opportunities to build a better city than the one that existed before.

Unfortunately, the movement (or relocation) of people away from Christchurch after the February 2011 earthquake, although inevitable, has potentially hindered the recovery process. Relocation is known to 'break up established neighbourhoods or communities and creates additional hardship for people, families and business, and impacts on schools, shops and community facilities. This means that recovery must focus on economic, social and cultural elements as well as the repair and rebuild of the built and natural environments' (CERA).

With this in mind, the long process of recovery is likely to be a difficult one.

Refer to the definition of a process above and **Resources 1A** and **1G – 1J** to answer the following questions.

a Evaluate the impact that the Christchurch rebuild is likely to have on:

i inward migration

ii the labour market.

ISBN: 9780170425285

Learning Activities

b Analyse the sequence of actions involved in the recovery of the Christchurch, Canterbury area following the 22 February 2011 earthquake.

ISBN: 9780170425285

The Canterbury region sits across the boundary of two tectonic plates, the Pacific Plate and the Australian Plate. Typical of landscapes located near convergent plate boundaries, Canterbury's landscape also shows evidence of significant folding (buckling) and faulting (fracturing) caused by the huge forces involved in plate movement. West of the Canterbury region lies the Southern Alps and Alpine Fault, a major fault that runs the entire length of the South Island along a SSW–NNE axis separating the Australian Plate to the west from the Pacific Plate to the east. The Pacific Plate is sliding against the Australian Plate at a rate of 35 mm and rising at a rate of 10 mm per year. To the east of the region are the Canterbury Plains. The Canterbury Plains are covered in young sediment (less than 2 million years in age) deposited as a result of extensive erosion of the Southern Alps, and transported by the Rakaia and Waimakariri rivers have carried huge amounts of gravel, sand and mud towards the eastern coast. These sediments have been deposited across the plain and exceed 200 m thickness in many places. Banks Peninsula southeast of Christchurch comprises basalt from volcanic activity more than 5 million years ago.

The 22 February 2011 earthquake fault runs beneath the north side of the Port Hills. The fault itself is not vertical. It slices through the ground on a 65-degree angle, back under the Port Hills, with the highest part of the fault lying approximately 1–2 km beneath the Avon-Heathcote Estuary. The two sides of the fault have moved past each other ('slipped') by about 1.5 m. The direction of the slip has caused the Port Hills to rise by about 40 cm.

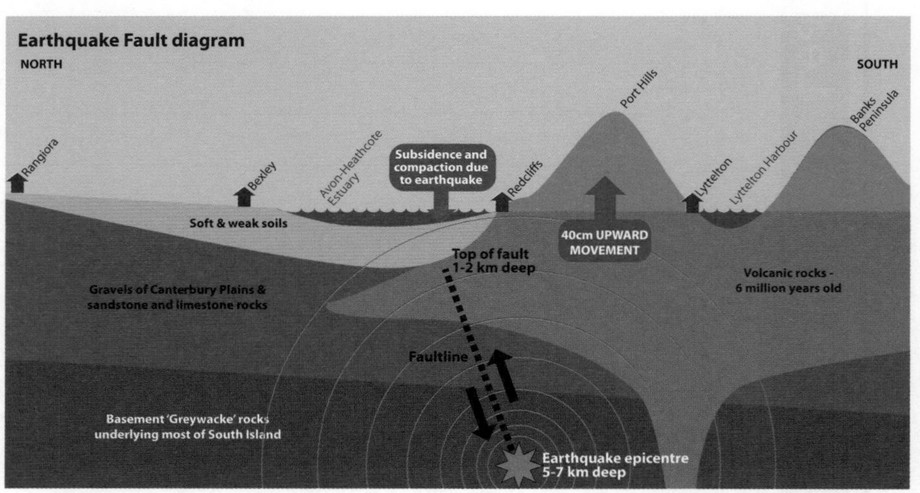

Resource 1A Geology of Christchurch, Canterbury

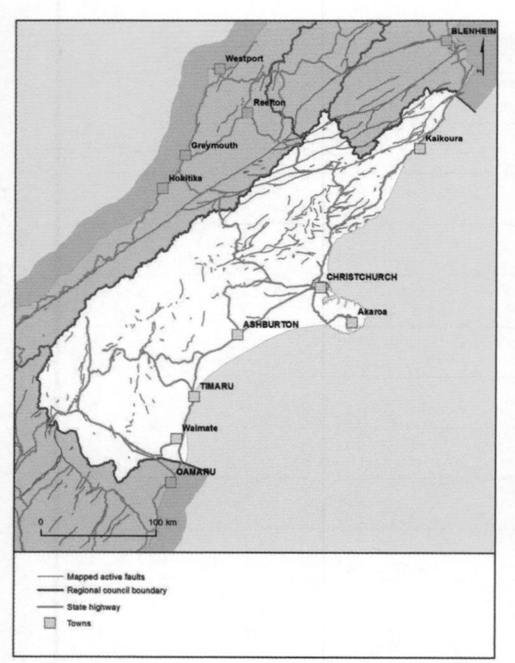

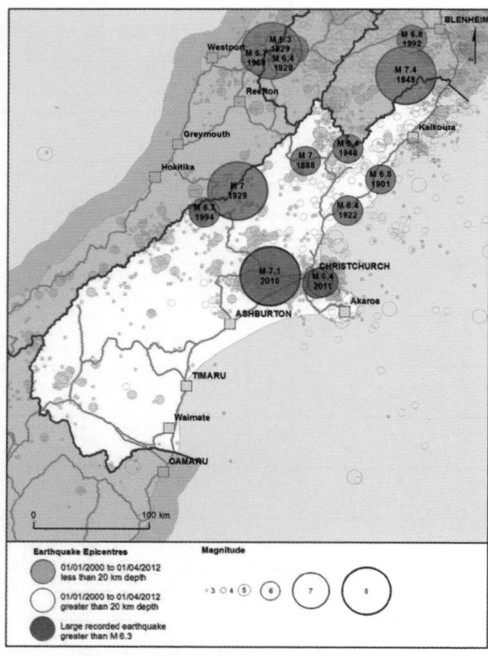

Resource 1B Mapped active faults in the Canterbury region (left) and earthquakes in Canterbury over the last 12 years (right)

The Canterbury plains are underlain by ancient faults that can be reactivated by accumulating stress at the margin of the Pacific-Australia plate boundary. In the central and northern South Island, because the crust of both the Pacific and Australian plates is very thick, the plates are colliding rather than subducting as is the case in the North Island.

About 75 percent of the movement between the Pacific and Australian plates is built up and released during major earthquakes along the Alpine Fault. East of the Alpine Fault, into Canterbury, the remaining 25 percent of the plate motion occurs through occasional earthquakes along a web of active faults. This motion extends all the way to the east coast, where faults such as those beneath the Canterbury Plains contribute 1–2 mm per year of the overall plate movement.

ISBN: 9780170425285

Region	Before 22 February 2011		After 22 February 2011	
	Number	%	Number	%
Northland	11	0.2	87	0.4
Auckland	218	3.5	985	4.1
Waikato	36	0.6	321	1.9
Bay of Plenty	35	0.6	240	1.0
Gisborne	0	0.0	12	0.0
Hawke's Bay	20	0.3	120	0.5
Manawatu-Wanganui	30	0.5	135	0.6
Wellington	124	2.0	440	1.8
Tasman	68	1.1	243	1.0
Nelson	70	1.1	225	1.1
Marlborough	46	0.7	222	0.9
West Coast	46	0.7	164	0.7
Canterbury	5237	82.9	19,742	81.4
Otago	167	2.6	802	3.3
Southland	36	0.6	166	0.7
Overseas	164	2.6	247	1.0

Resource 1C People relocations from the Canterbury region in the three months before and after 22 February 2011

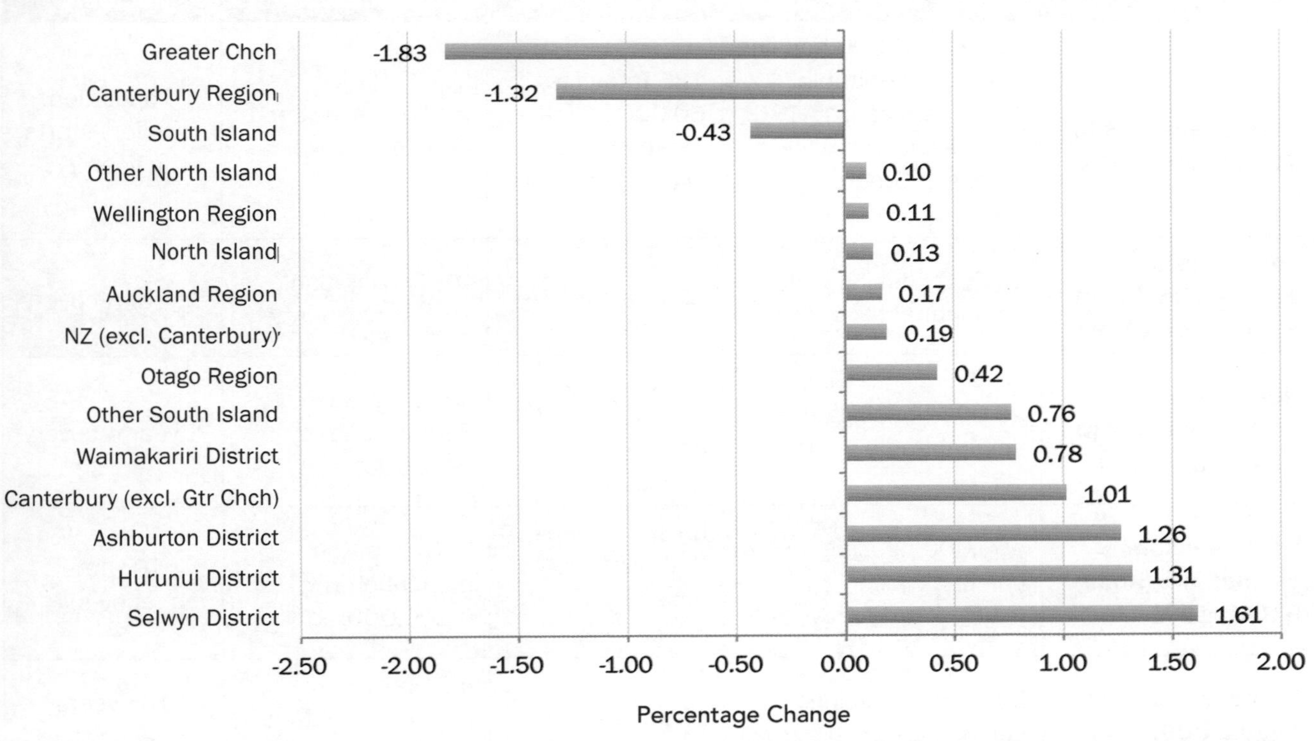

Resource 1D Short-term regional migration rates for 2010/2011

ISBN: 9780170425285

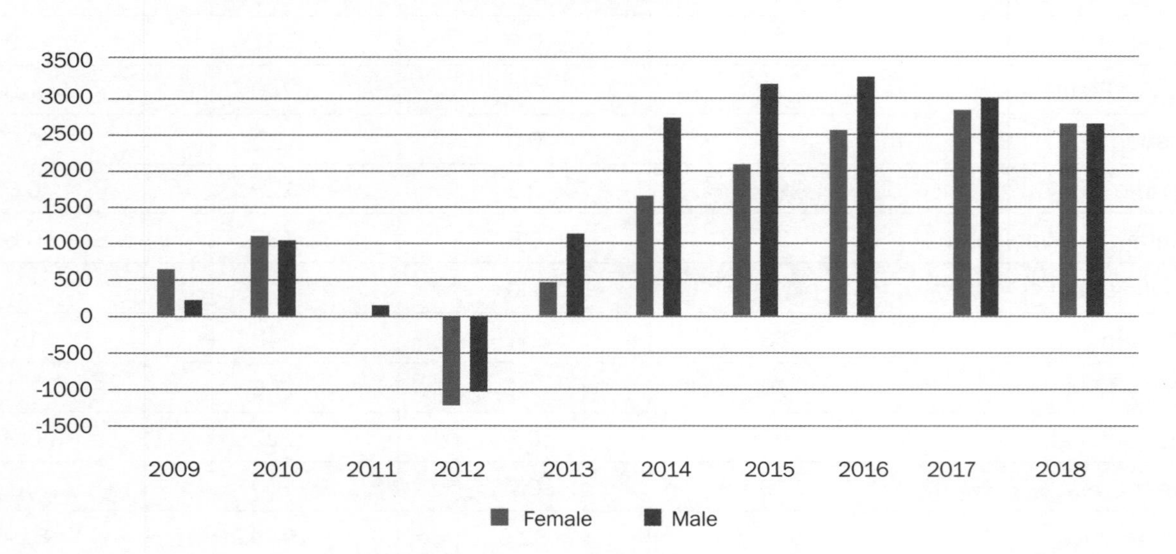

Resource 1E Christchurch city net permanent and long-term migration 2009–2018

Event	Building damage	Fatalities	Population effect scale	Population recovery
Christchurch earthquake, 6.3 M_L 22 February 2011	Red zone: 6000 housing units uninhabitable. CBD destroyed: 17,136. Jobs displaced: 9.8% of jobs in Greater Christchurch.	185 fatalities, 60,000 relocate for some one or two days?	2.4% drop in year 1 (8900)	Population recovery starts in year 2?
Loma Prieta earthquake, 6.9 M_L 17 October 1989	16,000 housing units uninhabitable	65 fatalities	0.3% fall in San Francisco County	1 year recovery
Northridge earthquake, 6.7 M_L 17 January 1994	25,000 housing units uninhabitable	60 fatalities, 30,000 emergency relocations	1% fall in San Fernando Valley (~12,000)	1 year recovery
Great Hanshin earthquake (Kobe), 6.3 M_w 1995	450,000 housing units uninhabitable, major port destroyed	6432 fatalities, 310,000 emergency relocations	6% fall in Kobe population (~100,000)	10 years recovery
Hurricane Katrina August 2005	200,000+ buildings uninhabitable	1836 fatalities	48% fall in population of New Orleans (~200,000)	10+ years recovery

Resource 1F Population impacts of selected hazard events

ISBN: 9780170425285

Resource 1G Evidence of Canterbury's progress in the recovery process (2018)

MINISTRY OF BUSINESS, INNOVATION & EMPLOYMENT
HIKINA WHAKATUTUKI

Labour Information

Search 

FAQs | Information campaigns | Contact us | News

Home | About | Employment Relations | Health and Safety | Immigration | Research | Key Projects | Publications | Careers

Home > Research Centre > Canterbury Labour Market research > Rolling Conversation 3

Research Centre
- Labour Market and Skills Research
- Canterbury Labour Market research
- Migration research
- Health & Safety research
- Employment Relations research
- Research Database
- About us

Canterbury rebuild - demand for occupational skills

Report No. 3, 12 December 2011

Summary

In the Department of Labour's (the Department) most recent contacts with construction companies involved with the Canterbury Rebuild, it was evident that there was very little change in the level of construction activity since September. The most prominent issues mentioned to the Department were:

- Companies have a reasonable flow of work but are unable to prepare properly for the expected rush of work later in 2012
- Companies had thought that the major buildup of work would start in the second quarter of 2012, but now expect the start to be delayed into the third quarter
- Many projects could be started if insurance was available. It was thought there may be some movement by insurance companies in February/March 2012 but no certainty of this happening
- Demolition, some new commercial building, residential and infrastructure repair work, is being carried out
- There are few concerns about labour availability at the moment. Companies have been hiring specialist staff from the United Kingdom and Ireland
- Although accommodation is currently available there is real concern for the future when it is expected that demand will significantly outstrip supply
- Companies are hoping that the Christchurch CBD Plan will clarify the nature and function of the CBD. Building costs may make it uneconomic for many previous tenants to return to the CBD
- The building consent process is being administered effectively by the Christchurch City Council but the likely volume of work later in 2012 will put the system under considerable pressure
- Immigration processing is efficient and companies are pleased with the service they have received to date. They would like Immigration New Zealand to clarify how the anticipated volume of applications is going to be handled when labour demand picks up.

Resource 1H Canterbury rebuild – demand for occupational skills

ISBN: 9780170425285

Stages of Recovery

Immediate	Short term	Medium to longer term
Provide basic human needs and support services.	Engage both established and new communities and inform them about rebuilding and future planning.	Continue to build resilient communities.
Address health and safety issues.	Establish new social and health support and service delivery models.	Major construction projects are under way.
Make safe or demolish unsafe and damaged buildings and structures.	Continue demolition of damaged buildings.	Complete restoration and adaptive reuse of heritage features.
Investigate, scope and initiate recovery programmes and initiatives.	Continue repair and rebuild.	Phase out recovery organisations.
Plan for land use and settlement patterns so land can be made available for displaced residents.	Deliver early projects to instil confidence.	Economy is growing and businesses are sustainable.
Conduct ongoing programme of investigation and research to understand the geotechnical issues and seismic conditions.	Planning and supporting community resilience.	Labour market is active and attracting employees.
Use this information to guide recovery activities and decisions on land suitability for rebuilding.	Begin replacement activity.	
	Begin restoration and adaptive reuse of heritage features.	
	Continue, monitor and review recovery.	

Resource 1I Stages of recovery

Building on the capacity, momentum and initiative of community-led responses to ensure the supportive networks in a community continue to thrive	Building on the strengths of the region, including clear roles and responsibilities that suit capabilities	Creating innovative solutions to past problems for a future-focused recovery
Two-way interaction and communication between all parties and better education about disaster responses	The importance of leadership, trust and transparency	Decision-making at the local level where possible
Focusing recovery work on the health and well-being of those people most affected	Economic recovery relies on retention of capital in the city and ability to retain financial equity	Restoring cultural, sporting and recreational life as part of community well-being, providing a sense of continuity with the past and a sense of shared identity
Government agencies working in a more innovative, flexible and collaborative manner, and in a more 'joined-up' approach with the private and volunteer sector	Coordinating recovery efforts and planning strategically for recovery and disaster resilience	Ensure future land-use decisions consider the seismically active environment and other natural hazards, such as those caused by building on land prone to liquefaction

Resource 1J Factors considered critical to a successful recovery (CERA)

ISBN: 9780170425285

The New Fossil Fuel Frontiers

Oil or petroleum is a fossil fuel that is sourced from deep beneath the ground. Its uses are numerous, providing fuel for transportation and machinery, to heating and energy for industry. However, as in other countries, New Zealand is consuming oil at an alarming rate and, as a result, is unable to produce enough oil to satisfy domestic demand. Consequently, New Zealand is becoming increasingly dependent on foreign oil supplies and now imports more than half of the oil it consumes, increasing its vulnerability to fluctuations in the international oil price. In 2010, in an effort to make New Zealand's energy needs more self-sufficient, the then National Government encouraged foreign oil companies to begin oil exploration beneath the large ocean basins northwest, south and east of New Zealand. In one such case, the Government of the time sold exploratory drilling permits for the Raukumara Basin to the Brazilian oil company Petrobras.

The Government's decision at the time was divisive in both political and public opinion, and a particularly controversial one for local iwi. It came as no surprise then when, in 2018, the newly elected Labour-led Government reversed the previous Government's decision and announced an immediate ban on the sale of new offshore oil and gas exploratory drilling permits, resurrecting the debate around the ethical importance of New Zealand's offshore oil and gas reserves.

Q1 Raukumara Basin

Learning Activities

Applying a geographic concept: Environments

Environments may be natural and/or cultural. They have particular characteristics and features, which can be the result of natural and/or cultural processes. The particular characteristics of an environment may be similar to and/or different from another.

For the most part, the seafloor of New Zealand's large offshore territory is uncharted. Nevertheless, data from recent reconnaissance surveys suggest that the large sedimentary basins that cover 20 percent of New Zealand's maritime territory (or approximately million sq km) may contain valuable reserves of oil and gas. Moreover, oil company interest in the Raukumara Basin stems from the basin's geology, which shares characteristics and features similar to that of other oil-producing basins around the world.

Refer to the definition of environments above and **Resources 2A – 2C** to answer the following questions.

a Describe in detail the Raukumara Basin according to its:

i absolute location

____

____

____

ISBN: 9780170425285

Learning Activities

ii size and extent.

b Analyse the Raukumara Basin according to its geological setting.

ISBN: 9780170425285

Q2 Environmental impacts

Applying a geographic concept: Interaction

Interaction involves elements of an environment affecting each other and being linked together. Interaction incorporates movement, flows, connections, links and interrelationships. Landscapes are the visible outcome of interactions. Interaction can bring about environmental change.

The perceived environmental risks associated with offshore oil exploration have been highlighted by events in New Zealand and overseas. The Deepwater Horizon oil spill in the Gulf of Mexico (2010) and the grounding of the *Rena* on Astrolabe Reef (2011) both demonstrated the effect oil contamination can have on the natural and/or cultural environment. Such risks need to be carefully weighed against the potential benefits of oil exploration on the New Zealand economy.

Refer to the definition of interaction above and **Resources 2D – 2G** to answer the following questions.

a Identify two positive impacts and two negative impacts oil exploration in the Raukumara Basin could have on local iwi Ngati Porou and Te Whanau a Apanui.

	Positive	**Negative**
First impact on local iwi		
Second impact on local iwi		

ISBN: 9780170425285

Learning Activities

b Discuss the potential effects of an oil spill on the coastal environment of Raukumara Peninsula.

ISBN: 9780170425285

Q3 Perspectives of exploratory drilling in the Raukumara Basin

Learning Activities

Applying a geographic concept: Perspectives

Perspectives are ways of seeing the world that help explain differences in decisions about, responses to, and interactions with environments. Perspectives are bodies of thought, theories or worldviews that shape people's values and have built up over time. They involve people's perceptions (how they view and interpret environments) and viewpoints (what they think) about geographic issues. Perceptions and viewpoints are influenced by people's values (deeply held beliefs about what is important or desirable).

The Government's decision to allow oil companies to begin exploratory drilling in the Raukumara Basin has not been without controversy. Since the Government decision there have been strong protests by environmental groups and local iwi. Nonetheless, plans to explore the Raukumara Basin for oil and gas remain on track.

Refer to the definition of perspectives above and **Resources 2D** and **2H – 2J** when answering this question.

a Choose two groups of people with different perspectives about exploratory drilling in the Raukumara Basin.

i Show their viewpoints about exploratory drilling on the continuum below.

Supportive of offshore exploratory drilling ⟷ Against offshore exploratory drilling

ii Justify why you have placed these two groups where you have in **i**. Use specific detail in your answer.

ISBN: 9780170425285

Learning Activities

ISBN: 9780170425285

Resources

The Raukumara Basin extends offshore from the northern coast of Raukumara Peninsula on the East Coast of New Zealand's North Island. The basin lies beneath the Raukumara Plain and is located between the volcanically active Havre Trough to the west and the Kermadec Trench plate boundary to the east. The basin's relative location is 250 km east of Auckland and about 125 km from the ports of Tauranga and Gisborne.

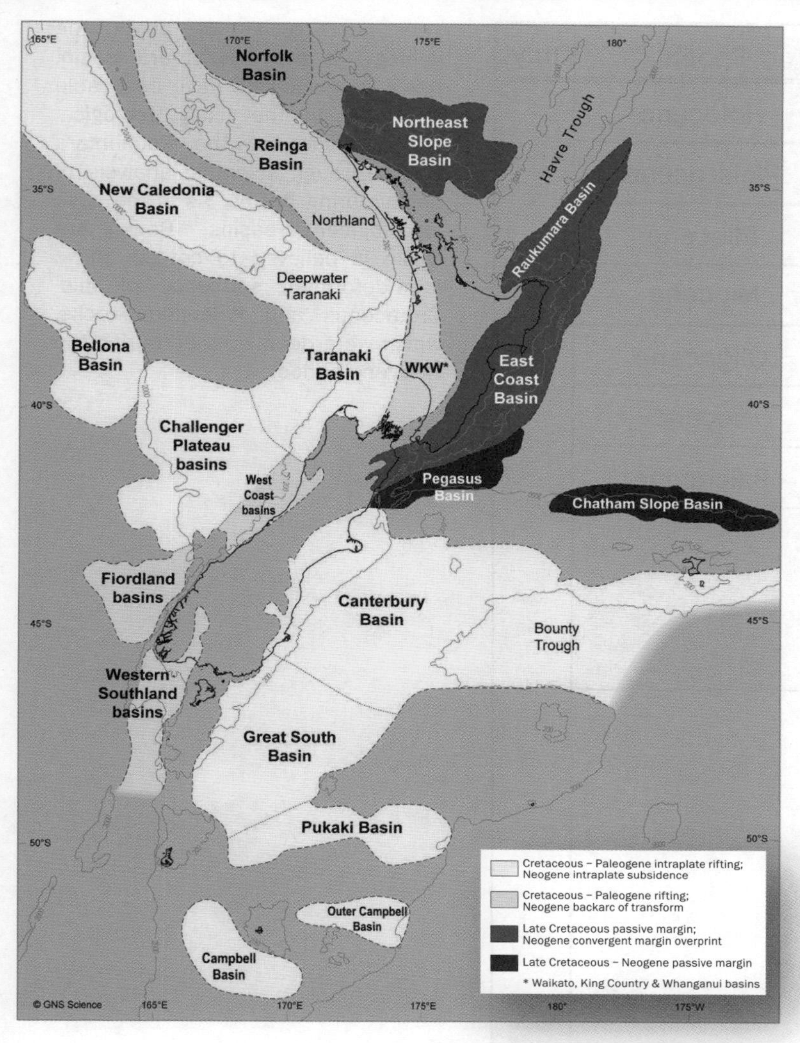

Resource 2A New Zealand's petroleum basins

The Raukumara shelf is about 10 km wide. From the shelf edge, the basin descends to around 2000 m and lies between 2000 and 3000 m below sea level. The sediments within the basin cover an area of about 30,000 sq km along a northeast axis, parallel to the East Coast continental margin. The northern boundary of the basin remains undefined.

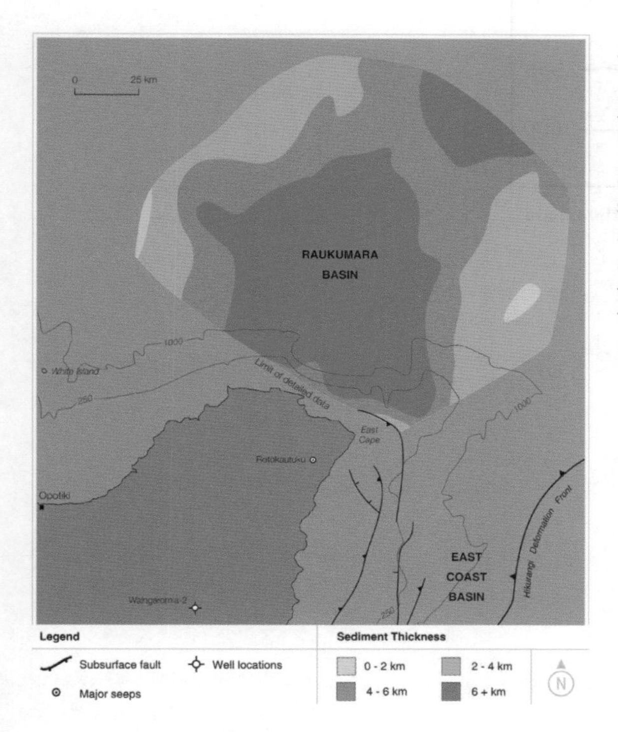

Resource 2B Size and extent of the Raukumara Basin

ISBN: 9780170425285

Eon	Era	Period	Epoch	m.y.
Phanerozoic	Cenozoic	Quaternary	Holocene	
			Pleistocene	1.5
		Neogene	Pliocene	
			Miocene	23
		Paleogene	Oligocene	
			Eocene	
			Paleocene	65
	Mesozoic	Cretaceous		
		Jurassic		
		Triassic		250
	Paleozoic	Permian		
		Carboniferous – Pennsylvanian		
		Carboniferous – Mississippian		
		Devonian		
		Silurian		
		Ordovician		
		Cambrian		540
Precambrian	Proterozoic			2500
	Archean			3800
	Hadean			4600

Resource 2C Geological timescale

Very little is known about the geology of the Raukumara Basin. However, geologists have been able to ascertain the likely geological characteristics of the Raukumara Basin through scientific analysis of neighbouring basins. As is the case in adjacent basins, Raukumara Basin is believed to contain thick, little-deformed, sequence of mid-Cretaceous and Paleogene rocks, which were deeply buried during the Neogene period.

ISBN: 9780170425285

Deep sea oil drilling in the Raukumara basin and Kaupapa Māori world views

'Mauriora' refers to the Maori cultural identity being secure (Durie, 1999). This security is of particular importance when considering access to 'te ao Maori'. 'Access' is therefore the fundamental concept to ensuring that mauriora is upheld. With regard to the advantages and disadvantages of Petrobras' plans, there are arguments from both sides. In a public address in 2010, then Energy and Resources Minister Gerry Brownlee suggested that an increased access to jobs, tax and royalty income, and regional development would come as a result of Petrobras' exploration (NZ Petroleum & Minerals, 2010a). This essentially means that some of the financial gains from Petrobras' exploration will be either directly or indirectly channelled into the local economy and infrastructure. Iwi members are particularly cautious, however, with the main concern being the practice of customary rights such as fishing and seafood collection potentially being impeded by seismic testing and drilling (Te Whanau a Apanui, 2011). Diminishing the integrity of the marine environment has the flow-on effect of diminishing access to Maori economic resources that come from the sea, in effect diminishing the cultural identity of local iwi who strongly identify with the sea.

'Waiora' refers to the physical environment being of a standard to promote, foster and sustain life (Durie, 1999). Waiora can therefore be directly applied to the effects on the physical environment that could propagate as a result of Petrobras' exploration. The advantages of undertaking exploratory spatial analysis and prospective drilling in the Raukumara Basin are very few when equating 'waiora' with a physical life-sustaining habitat for marine animals and fisheries. The fact that we have specific guidelines for the reduction of acoustic disturbance to marine mammals that must be met during seismic surveying implies that it, as a whole, is not beneficial to the marine ecosystem (Marine Conservation Unit, 2006). We have also recently witnessed the Deepwater Horizon oil spill in the Gulf of Mexico. Concordantly, the issue of deep-sea oil drilling is of great public concern. However, the definition of waiora could perhaps be framed in a more pro-exploration focus by defining the physical environment as an economic commodity. This approach would fit with the current government's stance on exploration, as a better understanding of the geology and petroleum-producing potential will effectively mean more financial leveraging with regard to mining the petroleum resource (NZ Petroleum & Minerals, 2010a). This could therefore increase the potential income stream created by selling petroleum stocks to increase the livelihoods of both local iwi and the rest of New Zealand. It must be acknowledged, however, that these projected opportunities are finite – due to the finite nature of the petroleum resource.

Resource 2D Advantages and disadvantages for local iwi Ngati Porou and Te Whanau a Apanui

ISBN: 9780170425285

On the evening of 20 April 2010, an explosion occurred on the Deepwater Horizon oil rig working on the Macondo exploration well for BP in the Gulf of Mexico. Eleven people died as a result of the explosion and many others were injured. The fire that followed burned for 36 hours before the rig sank. Following the sinking of the Deepwater Horizon oil rig, a seafloor oil gusher spilled oil for 87 days until it was finally capped on 15 July 2010. It is estimated that more than 4.9 million barrels of oil were discharged into the Gulf of Mexico between the time of the oil rig's initial explosion and sinking until the time the leak was capped, making it the largest accidental marine oil spill in history.

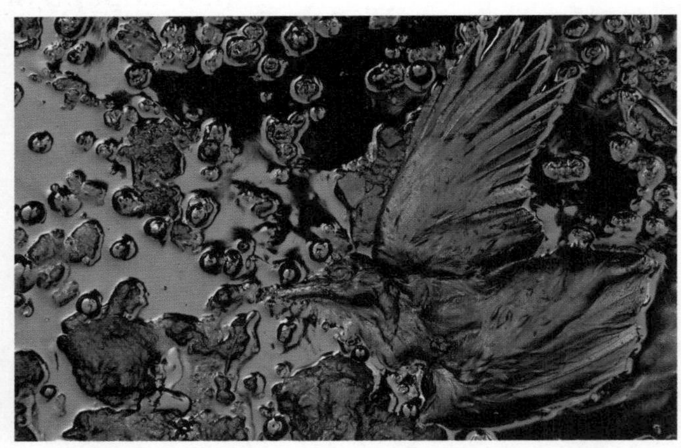

Resource 2E Deepwater Horizon

The impact of the Deepwater Horizon oil spill on the natural environment was considerable. However, the greatest impact was on the marine species. The area affected by the spill was inhabited by more than 4000 animal species including more than 1500 crustaceans, 1400 molluscs, 1200 fish, 200 birds, 29 marine mammals and 4 sea turtles.

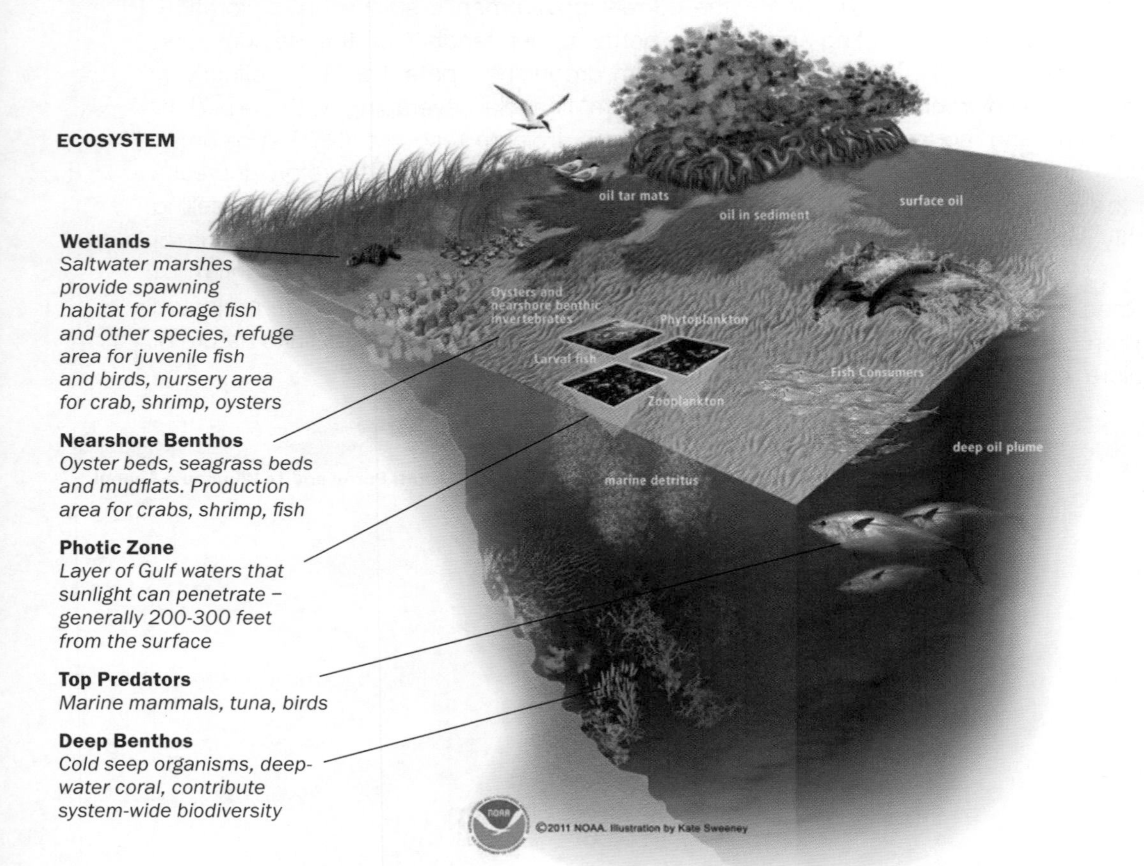

POTENTIAL OIL IMPACT

Wetlands
Oiled, degraded or eroding marsh may reduce productivity

Nearshore Benthos
Tar mats and oil in sediments may reduce benthic productivity and may affect food web

Photic Zone
Surface and dispersed oil may affect base of food web; larval fish are particularly sensitive to effects of oiling

Top Predators
Predators may be affected by degradation of food webs, and by direct health effects from oiling

Deep Benthos
Destruction of long-lived corals may reduce biodiversity and deep ocean productivity

Resource 2F Potential impact on marine environments

ISBN: 9780170425285

Rena incident overview

EARLY in the morning of 5 October 2011, the cargo vessel *Rena* struck Astrolabe Reef 12 nautical miles off Tauranga and grounded. The 21-year-old, 236-metre Liberian-flagged cargo vessel was en route from Napier to Tauranga and travelling at around 21 knots when it struck. Its bow section was wedged on the reef, and its stern section was afloat. Two of its cargo holds flooded and several breaches were identified in the hull. There were 25 crew on board *Rena* at time of grounding.

Oil spill response personnel and volunteers, including large numbers of locals, worked to clean oiled beaches and recover debris from the containers. Wildlife experts from the National Oiled Wildlife Response Team treated oiled birds, including little blue penguins and pied shags, and pre-emptively caught 60 rare New Zealand dotterel to prevent them becoming oiled. (These birds were later released back into cleaned environments in a staged release programme.)

Rena was carrying 1368 containers and 1733 tonnes of heavy fuel oil (HFO) on board at the time of grounding. An oil leak was detected on the night of 5 October and a salvor, Svitzer, was appointed by the vessel's owners and insurers the next day.

The salvage team began working around the clock in extremely dangerous conditions to secure the vessel and make preparations for the complex task of pumping the HFO off. The salvors began removing the estimated 1350 tonnes of oil in various tanks on *Rena* on 9 October, but were hampered by bad weather, equipment breakdown, and hazardous and changeable conditions.

A storm overnight on 11 October resulted in the loss of an estimated 350 tonnes of oil from *Rena*, some of it washing up at various points along the Bay of Plenty coastline. Continuing bad weather the following night saw 86 containers lost overboard. A further 5–10 tonnes of oil was lost from the vessel overnight on 22–23 October.

Over 1300 tonnes of HFO was eventually recovered from *Rena*, with all of the accessible oil removed by 15 November. Containers lost overboard during bad weather were intercepted and recovered, where possible, along with dispersed container contents that washed up. Container removal operations from *Rena* began once all of the oil had been removed, with the first container lifted off on 16 November 2011. By 26 December, a total of 341 containers had been removed.

Resource 2G *Rena* incident overview

ISBN: 9780170425285

The New Zealand Herald

Editorial: Give deep-sea oil drilling a fair chance

NEW ZEALAND as geologists know it is many times larger than the map we recognise. The land above water is just the highest part of a crustal block extending far off the coast, mainly west and north of the North Island and east and south of the South Island.

These submarine plateaus may be as valuable to our economy as anything on or under dry land. The fishing grounds of the Chatham Rise and the gas wells off Taranaki have already contributed a great deal. There is bound to be more wealth around us.

But do we want to find it? That is a question we have to resolve. The Government has suffered two setbacks in attempts to explore the country's mineral deposits. A public outcry and protest marches greeted its proposal for prospecting in a few promising sites on the conservation estate, and last year East Cape Maori made common cause with Greenpeace to stop the Brazilian oil producer Petrobras drilling in the Raukumara Basin.

It might not be clear to all the protesters that these ventures are exploratory. There is no certainty that minerals in commercial quantities will be found and if they are, no guarantee that they can be economically extracted, and at cost that is competitive with deposits of the same mineral in other parts of the world.

Once a prospect passed all those tests the Government would then need to convince New Zealanders the value of mining would outweigh any environmental damage.

The last test could be the most difficult of all. If mere exploration can trigger the scale of protest that forced the Government to back down over conservation sites, a mining application might stand no chance. Nevertheless, we should make that decision with the best knowledge of what is down there.

The Green Party, which has led much of the opposition to "mining", held its annual conference at the weekend and seems to have adopted a more reasonable attitude to the issue. Co-leader Russel Norman said the party was not opposed to all mining, only to the extraction of coal and oil, which in its view have no future, and processes such as deep-sea drilling and fracking, which it considers too risky.

Buoyed by their election result last year, the Greens are aiming to be a partner in the next Labour government. While Labour does not share the Greens' aversion to more coal mining, Dr Norman said that was a difference that could be discussed if a Labour-Greens coalition was in prospect. Hopefully deep-sea drilling would be discussed in the same spirit if a well of oil or gas has been discovered by then.

New Zealand is getting a more balanced environmental law for ocean drilling. The economic benefits of extraction will be no less important than the marine ecosystem when a project is considered. The in-built environmental bias of the Resource Management Act will continue to restrict development on land and within the 12-mile jurisdiction. But the rest of New Zealand's exclusive economic zone will be easier to exploit, at least legally. Physically is another question. The Southern Ocean seabed is ridged and trenched and turbulent.

Offshore oil drilling is haunted by the accident in the Gulf of Mexico two years ago that caused one of the world's worst oil spills. The well gushed for three months while efforts were made to plug it. A rare accident such as that is a lesson for all concerned, it is not an indictment of the industry unless it happens too often.

New Zealand needs all the mineral wealth it can find. It cannot sacrifice it all for a pristine environment on land, still less beneath the sea. It is a matter of striking a reasonable balance. Ocean prospecting is expensive, internationally competitive and not often rewarded. The law should not add unnecessarily to the odds against success.

Resource 2H Editorial: Give deep-sea oil drilling a fair chance

ISBN: 9780170425285

THE OVERARCHING GOAL OF THE GOVERNMENT is to grow the New Zealand economy to deliver greater prosperity, security and opportunities for all New Zealanders. New Zealand is blessed with extraordinary energy resources, which have the potential to make a significant contribution to our prosperity and our economic development.

Most New Zealanders know that we have an abundance of renewable energy resources. We continue to be a world leader in geothermal energy. Our rivers and lakes have long provided clean hydro-electricity. Our wind resources are world class. New Zealanders are exploring how to harness the waves, the tides, and the sun in order to generate power.

What is less well known is that along with our renewable resources, we also have an abundance of petroleum and mineral resources. More than 1.2 million square kilometres of our exclusive economic zone are likely to be underlain by sedimentary basins thick enough to generate petroleum. Recent reports put New Zealand's mineral and coal endowment in the hundreds of billions of dollars. For too long now we have not made the most of the wealth hidden in our hills, under the ground, and in our oceans. It is a priority of this government to responsibly develop those resources. The New Zealand Energy Strategy sets the strategic direction for the energy sector and the role energy will play in the New Zealand economy. The government's goal is for the energy sector to maximise its contribution to economic growth. The Strategy focuses on four priorities to achieve that: developing resources; promoting energy security and affordability; achieving efficient use of energy; and environmental responsibility.

This New Zealand Energy Strategy includes the New Zealand Energy Efficiency and Conservation Strategy. Energy conservation and efficiency has an important role to play in economic growth. All New Zealanders benefit from more effective use of our resources. The Energy Efficiency and Conservation Strategy is therefore all about practical actions that encourage consumers of energy to make wise decisions and choose efficient products.

I am confident these strategies provide the government's vision for New Zealand's energy sector – one that is efficient and contributes to the economic prosperity of all New Zealanders.

Hon. Hekia Parata
Acting Minister of Energy and Resources

Resource 2I Government: Developing our energy potential

ISBN: 9780170425285

Protest flotilla heads out to confront seismic oil detection ship

Opotiki, Monday, 4 April 2011: The flotilla opposed to deep-sea oil drilling entered the zone in the Raukumara Basin where seismic testing is scheduled to begin today.

After a powerful welcome and meeting with local iwi Te Whanau a Apanui on the weekend, skippers of the Stop Deep Sea Oil flotilla resolved to sail out and meet the seismic testing vessel in the protest tradition of 'bearing witness' used during the decades of the Nuclear Free Pacific campaign. A boat from Te Whanau a Apanui has joined the flotilla.

The flotilla's largest sailing vessel, *Infinity*, departed Whangaparaoa, East Cape, North Island yesterday and sailed through the night to enter the testing zone early this morning. It is expected to encounter the *Orient Explorer* in the next 24 hours. Greenpeace New Zealand spokesperson Steve Abel said, 'As New Zealanders we regard our coastlines and oceans as national treasures that are much too valuable to risk with oil spills.' The flotilla's departure coincides with the leaking of the Government's Energy Strategy.

'Right as the Deep Sea Oil protest flotilla is fighting for a clean energy future and is determined that deep-sea oil drilling does not happen in New Zealand waters, this current Energy Strategy will mire New Zealand into an even deeper dependency on polluting fossil fuels. The Government couldn't have got it more wrong,' said Steve Abel. 'No one is saying petrol pumps will be turned off tomorrow. Our dependence on fossil fuels won't end overnight but our investment from now on needs to be in clean energy – not looking for the last drops of oil in the most risky places.'

The *Orient Explorer* is under contract to Brazilian petrol giant Petrobras to do seismic surveying. The permit was granted by Gerry Brownlee to Petrobras in 2010, at the same time as oil was pouring into the US Gulf of Mexico in the infamous BP deep-sea oil disaster.

To estimate the size of an oil reserve beneath the deep sea floor, survey ships tow up to 10 kilometres of multiple airgun floats that emit thousands of high-decibel explosive impulses to map the geology beneath the sea floor. Seismic surveys have been implicated in harming marine life and migrations, including whale beaching and stranding incidents. The Department of Conservation states that beaked whales live in the Raukumara Basin. The flotilla is opposing all aspects of the deep-sea exploration and drilling programme.

'Just as Energy Minister Gerry Brownlee misled the mining industry last year by attempting to open up high-value conservation land to mining, our government has misled the deep-sea oil industry by saying our oceans are open to them. Last year New Zealanders opposed mining on our conservation land and the Government listened. The Government now needs to respond to the growing opposition against deep-sea oil drilling by stating that deep-sea oil companies are not welcome here,' concluded Abel.

Resource 2J Greenpeace: Protest flotilla heads out to confront seismic oil detection ship

ISBN: 9780170425285

Shanghai Sprawl

The city of Shanghai sits along the delta banks of the Yangtze River on the eastern coast of China. The 2010 census put Shanghai's total population at 23 million people, a growth of 37.5 percent from 16.7 million in 2000 making it the world's most populous city proper and one of its fastest growing. As economic development and modern agricultural practices reduce the need for a large rural agricultural labour force, mass migration from the countryside accounts for the majority of this recent growth.

Nevertheless, Shanghai's growth has come at a considerable cost. Urban sprawl has put a strain on the city's limited resources. Water availability is a key concern, and excessive groundwater extraction has resulted in extensive subsidence in and around the city. Because the city's urban area sits alongside the Yangtze River, just a few metres above sea level, subsidence is a critical concern.

Q1 Shanghai: site and situation

Learning Activities

Applying a geographic concept: Environments

Environments may be natural and/or cultural. They have particular characteristics and features, which can be the result of natural and/or cultural processes. The particular characteristics of an environment may be similar to and/or different from another.

Located on the Yangtze River delta, Shanghai is well positioned for international trade because of its location and proximity to the Pacific Ocean and East China Sea. The inner city of Shanghai surrounds the Huangpu River, a tributary that leads to the Yangtze River north of the city. Subsequently, most of urban Shanghai is built on flat alluvial plains.

Refer to the definition of environments above and **Resources 3A** – **3E** to answer the following questions.

a Accurately describe the absolute location of Shanghai using latitude and longitude (degrees and minutes).

b Complete the précis map on the following page by naming and indicating the location of the following features:

- **i** Hydrological features and the direction of flow
- **ii** The Bund and Oriental Pearl Tower east of the Huangpu River shown in **Resource 3B**
- **iii** Areas of recreation north of Suzhou Creek
- **iv** Transport features.

ISBN: 9780170425285

Learning Activities

Key

ISBN: 9780170425285

Learning Activities

c State the compass direction the photographer of **Resource 3B** was facing when the photograph was taken.

d The natural environment of Shanghai can be perceived as both an advantage and a disadvantage for its economic growth. Give an example of each, using evidence from the resources.

i One advantage for economic growth, with evidence.

ii One disadvantage for economic growth, with evidence.

iii Discuss whether Shanghai's natural environment promotes or impedes economic development.

Q2 Rapid population growth and urbanisation

Applying a geographic concept: Change

Change involves any alteration to the natural or cultural environment. Change can be spatial and/or temporal. Change is a normal process in both natural and cultural environments. It occurs at varying rates, at different times and in different places. Some changes are predictable, recurrent or cyclic, while others are unpredictable or erratic. Change can bring about further change.

Of the 23 million residents of Shanghai, about 9 million, or more than 39 percent, are unregistered long-term migrants. The main origins of the migrants are Anhui (29.0 percent), Jiangsu (16.8 percent), Henan (8.7 percent), and Sichuan (7.0 percent) provinces. More than 79 percent of internal migrants are from rural areas. Internal migrants account for the entire population increase, as Shanghai's natural population growth rate has been negative since 1993 due to its extremely low fertility rate (0.6), which came about in response to China's One Child Policy.

a Refer to the text above to answer this question.

i Use an appropriate statistical mapping technique to display the magnitude and origin of migrants who moved to Shanghai as unregistered migrants. Use appropriate mapping conventions.

ISBN: 9780170425285

ii Based on the graphical map you have drawn, explain in detail the pattern of internal migrants to Shanghai. Tip: use PQE.

b Refer to **Resource 3F** when answering these questions.

i Draw an appropriate graph below, using the data provided in **Resource 3F**, which shows changes in Shanghai's population size structure from 1953 to 2010. Observe all conventions in the construction of your graph. Tip: use SALTS.

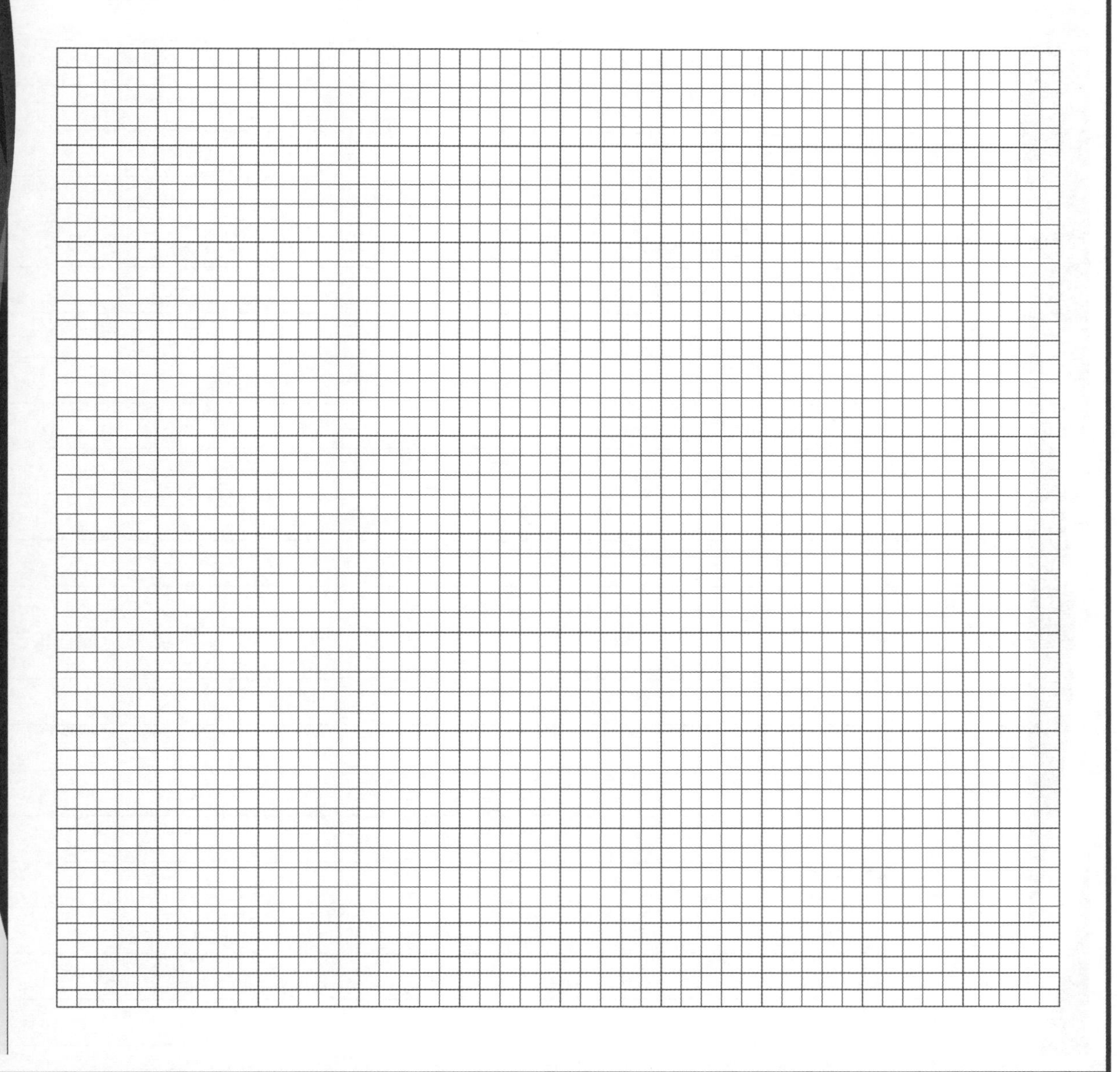

ISBN: 9780170425285

Learning Activities

ii Describe the change in population structure from 1953 to 2010.

iii Account for the shape of Shanghai's age-sex structure shown in **Resource 3G**.

ISBN: 9780170425285

Q3 Sustainability

Applying a geographic concept: Sustainability

Sustainability involves adopting ways of thinking and behaving that allow individuals, groups and societies to meet their needs and aspirations without preventing future generations from meeting theirs. Sustainable interaction with the environment may be achieved by preventing, limiting, minimising or correcting environmental damage to water, air and soil, as well as considering ecosystems and problems related to waste, noise, and visual pollution.

As has happened throughout China over the past three decades, urbanisation in Shanghai continues to progress at a relentless pace. Left unchecked, urbanisation in the long term will lead to urban sprawl resulting in unsustainable use of marginal lands.

Refer to the definition of sustainability above and **Resources 3H** and **3I** to answer this question.

a Comprehensively analyse the problems associated with urbanisation in Shanghai giving reasons for the changing demand for land around Shanghai's urban-rural fringe.

ISBN: 9780170425285

Learning Activities

ISBN: 9780170425285

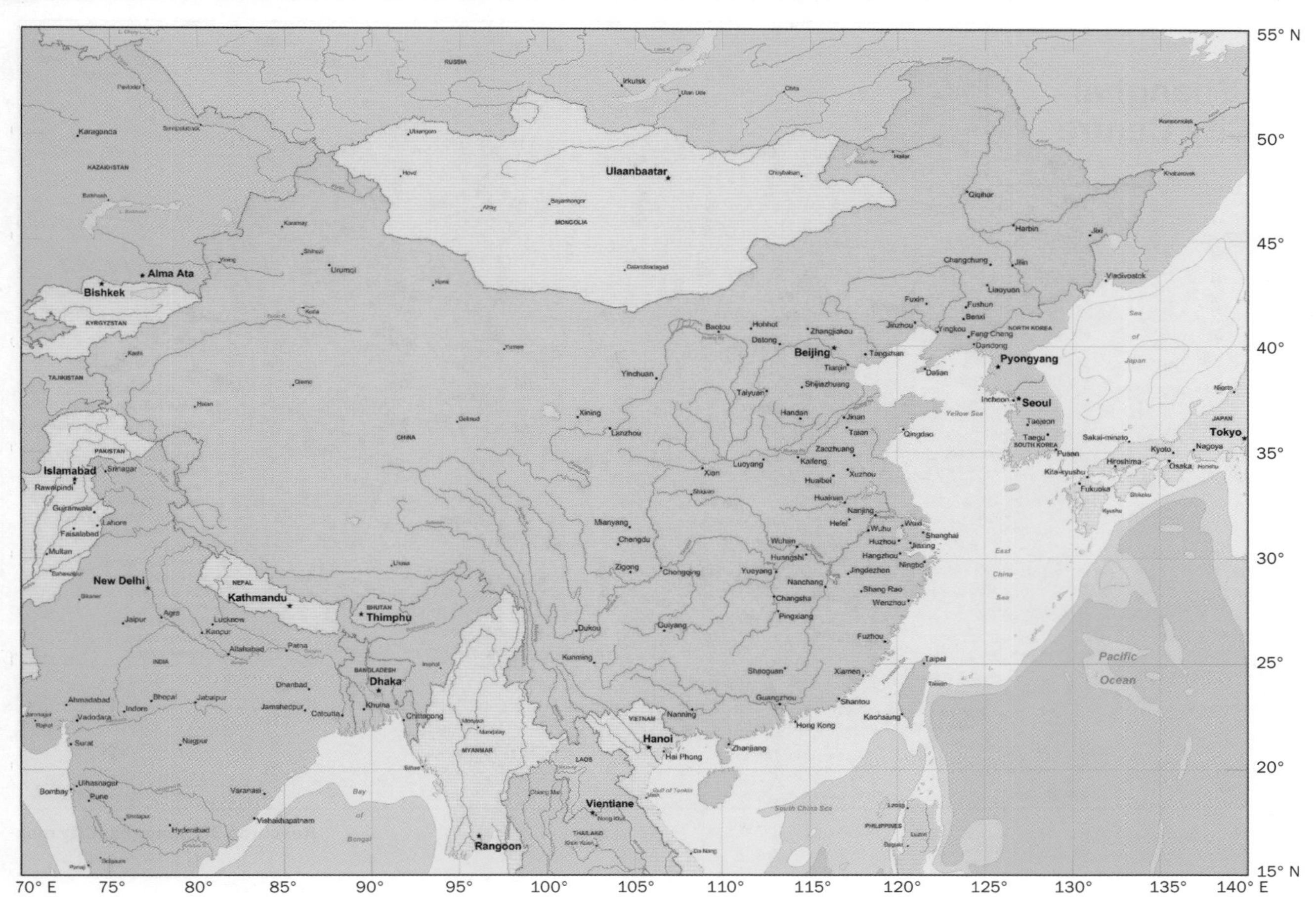

Resource 3A China location map

Resource 3B The Bund (right) and the Oriental Pearl Tower (background left)

ISBN: 9780170425285

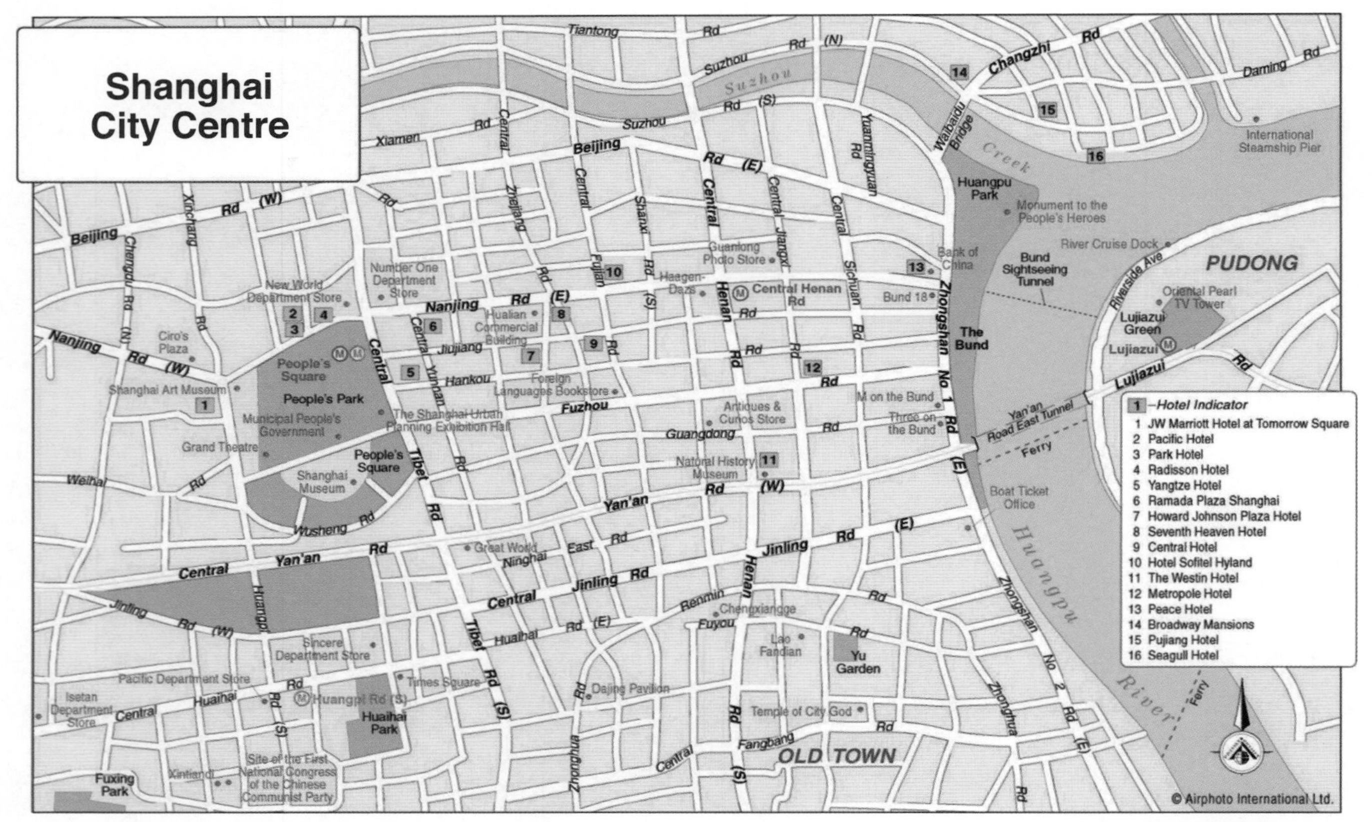

Resource 3C Shanghai City map

Resource 3D Container terminal located on the Huangpu River, Shanghai

Resource 3E Water village Zhouzhuang, Shanghai

ISBN: 9780170425285

Population by age group	First census (1953)	Second census (1964)	Third census (1982)	Fourth census (1990)	Fifth census (2000)	Sixth census (2010)
Aged 0–14	2,048,900	4,576,100	2,154,200	2,431,800	2,010,900	1,982,900
Aged 15–59	3,927,400	5,583,100	8,340,200	9,018,900	11,939,200	17,566,600
Aged 60 and above	228,100	657,300	1,365,300	1,891,200	2,457,600	3,469,700
Aged 100 and above	10,000	20,000	200,000	800,000	2,690,000	9,280,000

Resource 3F Inter-census demographic change

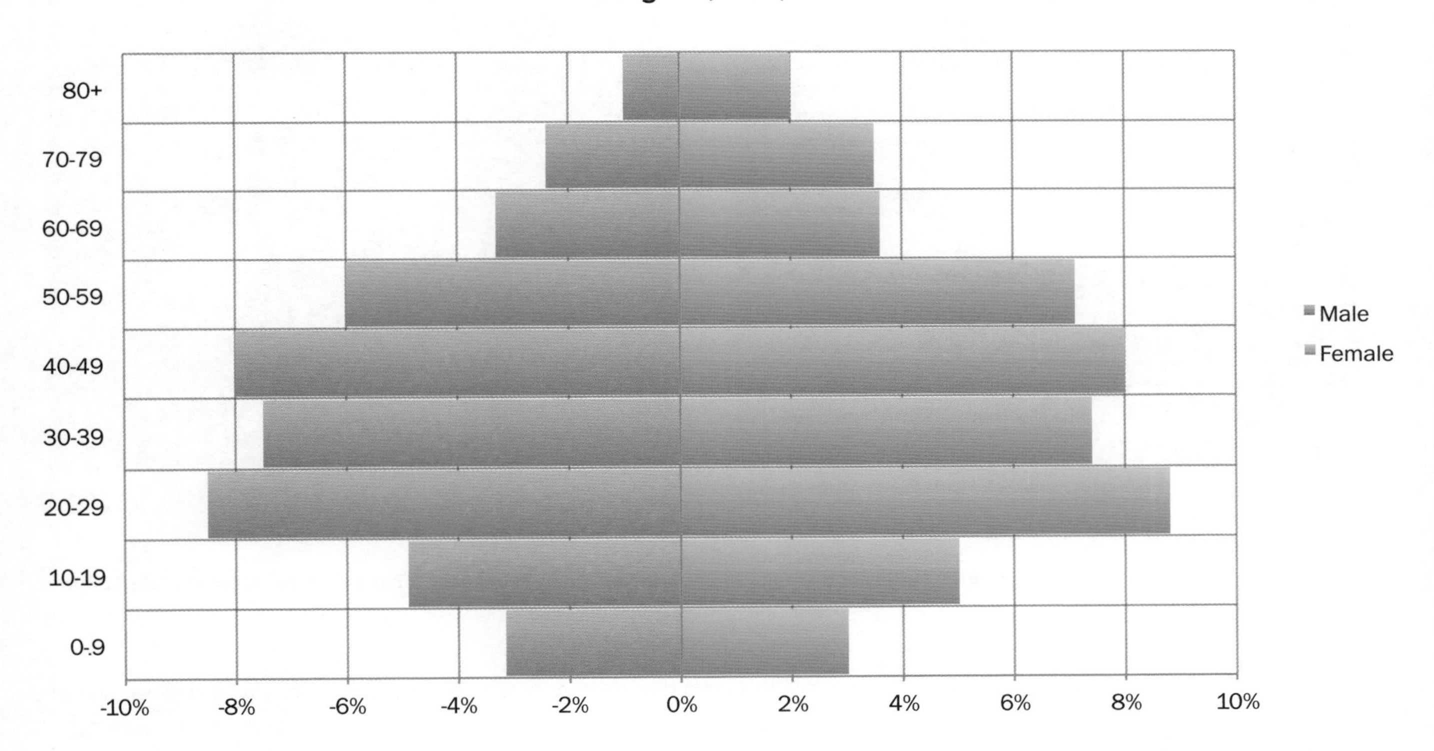

Resource 3G Age-sex structure for Shanghai

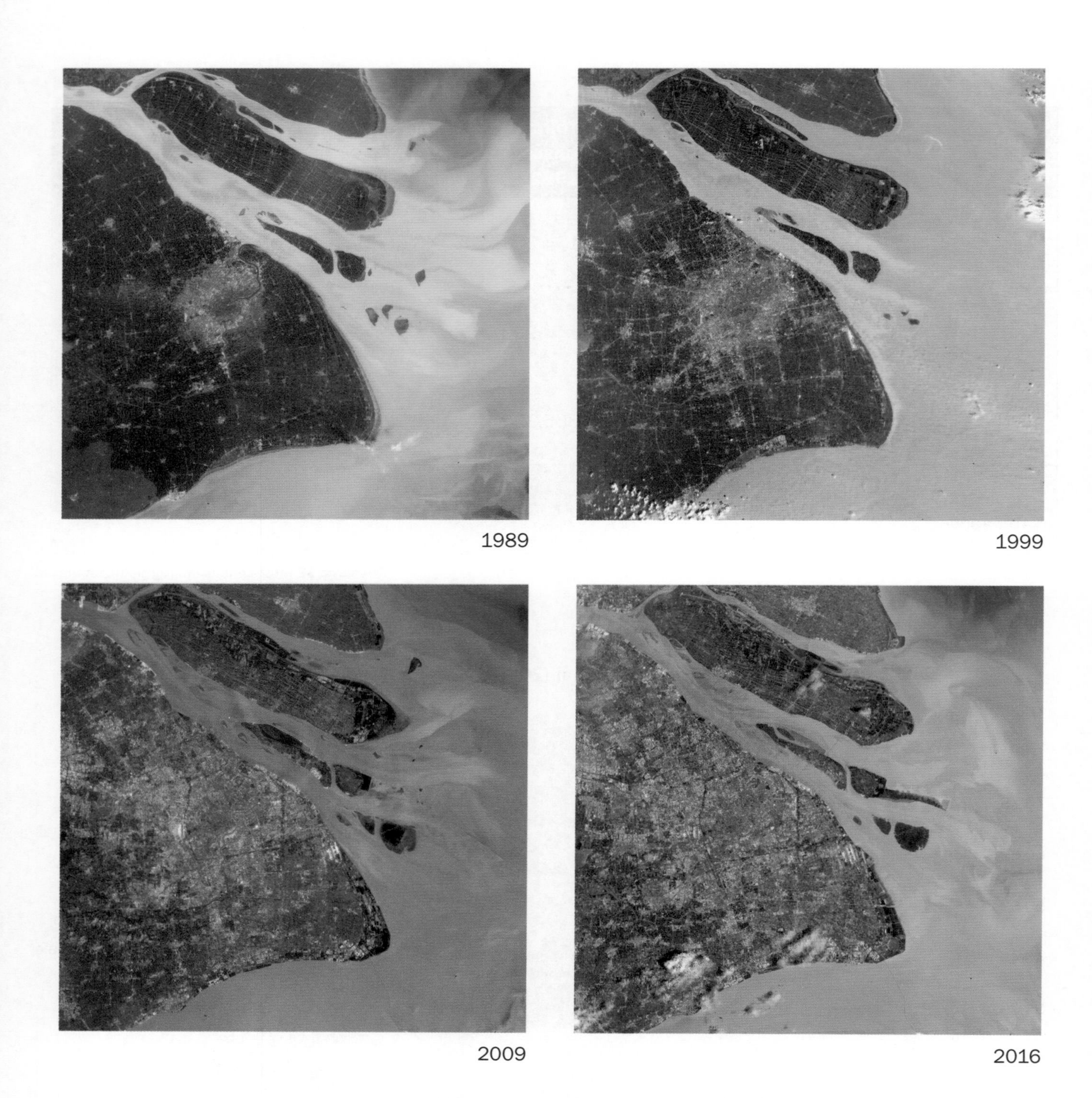

Resource 3H Aerial photographs showing the extent of Shanghai's urban expansion since the late 1980s.

If you could go back in time to the 1980s, you would find a city that is very different from today's Shanghai. What had been a relatively compact industrial city of 13 million people in 1989 swelled to 24 million in 2016, making it one of the largest metropolitan areas in the world. The city's rapid expansion and associated urban sprawl have not only changed patterns of land use within Shanghai, but also patterns of land use in the surrounding Yangtze River Delta.

ISBN: 9780170425285

Soaring to Sinking: How Building Up Is Bringing Shanghai Down

As land-subsidence concerns sweep across more than 50 cities in China, the country's most populous metropolis remains among the most vulnerable

Shanghai's skyline glitters with sleek financial skyscrapers and glossy residential towers, but below the city's lustrous facade lies an enduring problem. Thanks to mass urban migration, soft soil and global warming, Shanghai is sinking, and has been for decades. Since 1921, China's most populous city has descended more than 6 ft. Across China, land subsidence affects more than 50 cities, where 49,000 sq. mi. of land have dropped at least 8 in.

It's not just the numbers that are frightening: the problem has manifested itself tragically and more frequently of late. Earlier this month, a young woman unexpectedly fell through the sidewalk into a 20-ft.-deep sinkhole while walking along the street in Xi'an. In April, a woman died after falling through the sidewalk into a pit of boiling water in Beijing. Scientists have continuously warned of dire repercussions if the government does not implement more stringent guidelines for urban planning, water usage and carbon emissions — and they expect the situation to get much worse in areas with large-scale, fast-paced construction, like Shanghai.

As progress continues on Asia's soon-to-be-tallest skyscraper, the Shanghai Tower, the problem has manifested itself in malicious cracks nearby, captured and posted by users of the Twitteresque microblogging site Sina Weibo, then published by ChinaSmack. In mid-February, one blogger posted about a 22-ft.-long crack situated near the 101-story Shanghai World Financial Center, across the street from the highly anticipated tower.

In response to bloggers' concerns, Shanghai Tower Construction, the company responsible for building the tower, issued a statement saying surface cracks were "controlled and safe," the Shanghaiist reported. Liu Dongwei, chief architect of the China Institute of Building Standard Design & Research, cited groundwater, rainfall and soft soil foundation as the reasons for the settlements. But that's only partially accurate.

Shanghai has inherently soft soil because of its geographical position at the mouth of the Yangtze River basin and, yes, groundwater accounts for nearly 70% of land subsidence; however, experts say, the weight of skyscrapers and global warming also play hefty roles in aggravating what they call "the most important geological disaster in Shanghai." Unfortunately, the implications will only grow graver with the pace of development and rising sea levels.

According to a report by the Shanghai Geological Research Institute, the physical weight of skyscrapers accounts for 30% of Shanghai's surface subsidence. "Usually groundwater pumping is the key factor," Jimmy Jiao, a professor of earth sciences at Hong Kong University, tells TIME. "But in Shanghai, development is also important because the building density is high, and most of the high-rise buildings are sitting on the areas with soft soil." Basically, what's happening is, the weight of high-rise towers presses down on the earth, as if you were to put a weight on a spring or scale.

The most densely packed city in China, Shanghai reached a population of 23 million in 2010, according to census data. With the soaring number of denizens flocking to the seaside city, developers have stacked skyscrapers and thousands of high-rises side by side like dominoes. The country as a whole built 200 new skyscrapers in 2011, and by 2016, the total number is expected to exceed 800.

"As the saying goes, the more you build, the more they come," Jiang Li, a professor of civil engineering at Baltimore's Morgan State University who grew up in Tianjin, China, tells TIME. Pretty soon there will be 30 million people in Shanghai, while Beijing is just short of 20 million people. "At this point, so much construction has already been done that the areas are hopeless," he says. Today's problems have been aggravated

(continued over page)

Resource 3I Soaring to sinking: how building up is bringing Shanghai down

by decades of overdevelopment and overpumping of groundwater resources.

The problems began in the 19th century, when Shanghai transformed into a trading port and began attracting both foreigners and relocating Chinese migrants. By 1900, the population had tripled to more than 1 million. People started consuming more groundwater than the overlying turf could handle, and the problem worsened dramatically. By the 1950s and early '60s, the area started sinking 4 in. per year. The pace slowed after 1963, when the government banned a significant number of wells. To take further precaution, the government also began pumping water back into underground reservoirs. Every day, Shanghai is redirecting 60,000 tons of water through 121 wells, *China Daily* reported. Even with these restrictions in place, the city has descended 16 in. in the last 50 years.

Shanghai may have had this problem before the 1950s, but it didn't start emerging in other cities until the early '80s. Now more than 50 cities across the country face sinking problems, according to a report by the China Geology Survey. Three regions in particular have "serious land-subsidence problems," including the Yangtze River delta area, the Fenhe River–Weihe River basin and the North China Plain. According to CCTV, Cangzhou, a city in north China's Hebei province, has descended nearly 7 ft. In 2009, the city had to demolish a three-story building housing a branch of the city's People's Hospital because the first level sank so low that it fell underground.

Though some critics argue that the Chinese government has been too slow to act, research, public concern and some hefty bills ($35 billion in Shanghai alone in the past 40 years) have sparked some momentum. Recently the state council approved China's Land Subsidence Prevention Project, a countrywide initiative to prevent land subsidence. Likewise, Beijing, which has descended more than a foot in the past decade, has also made an effort to reduce underground-water extraction, with plans to close 800 water-extraction wells this year, according to the Beijing Water Authority. By 2014, the city hopes to halt underground-water extraction in urban areas altogether as part of the North-South Water Diversion Project. The project expects to deliver 3 billion cu. ft. of water supply to Beijing from the Yangtze River. This would not only satisfy one-third of the city's total water demand but would also cut the extraction of underground water in half.

But Li, who worked at the Chinese Academy of Science for 15 years, says such programs will not be enough. "It's hard to quantify how much this might help, but the question is, Is that a problem solved? The answer is no. The problem lies in the early issue with urbanization," he says. Scientists expect the regulations to help curb the consumption of underground-water supplies, but there are a few things the government has less control over, like global warming. As land degradation and excessive guzzling of groundwater continue, environmentalists predict waters surrounding Shanghai will rise 9 to 27 in. by 2050 as a result of melting ice caps.

"If you look at Shanghai during high tide, you can see the water level is higher than the streets but separated by the wall," Li says. "This is a situation where if you have a major disaster like a hurricane, tsunami or tropical storm, it can cause serious damage." He is especially worried about severe flooding in the coastal areas, where the majority of Chinese migrants have settled. The only way to really solve the problem is to reduce — or better yet, stop — groundwater pumping. Another option is to decrease the density of buildings, which would mean fewer heavy skyscrapers, perhaps an unrealistic solution for China's rapidly growing cities.

In the meantime, Li suggests local governments impose water restrictions and fees to encourage less wasteful consumption of water. He also proposes more secondary water uses, where wastewater is recycled for washing cars or watering plants. Even then, global warming remains an obstacle. As skyscrapers in Shanghai go up and the glaciers in the North and South poles melt down, cities like Shanghai grow more and more vulnerable every day.

ISBN: 9780170425285

Nuclear Meltdown

Only twice in human history has a nuclear incident been awarded the highest rating of 7 on the International Nuclear and Radiological Scale. In the first event, in 1986, the Chernobyl nuclear power station in Ukraine (then part of the former Soviet Union) exploded, destroying most of the station's reactor building. The explosion was the consequence of direct human error. The main disaster, however, came not from the explosion but from the widespread fallout of radiation that followed. The second incident occurred near Fukushima, Japan, in 2011. Unlike the Chernobyl disaster, the initial cause of the Fukushima Daiichi nuclear meltdown was very different, in that the sequence of events that led to the reactor meltdown and subsequent release of radioactive material began with a devastating magnitude 9 earthquake that generated a powerful tsunami.

Q1 Processes change environments

Learning Activities

Applying a geographic concept: Processes

A process is sequence of actions, natural and/or cultural, that shape and change environments, places and societies. Some examples of geographic processes include erosion, migration, desertification and globalisation.

The sequence of actions that led to the Chernobyl and Fukushima Daiichi disasters, although similar in impact, contrast starkly.

Refer to the definition of a process above and **Resources 4A – 4E** to answer the following questions.

a To what extent should the Chernobyl and Fukushima Daiichi disasters be considered technological hazard events?

Learning Activities

b With reference to **Resource 4A**, describe the sequence of actions that preceded the Fukushima Daiichi incident.

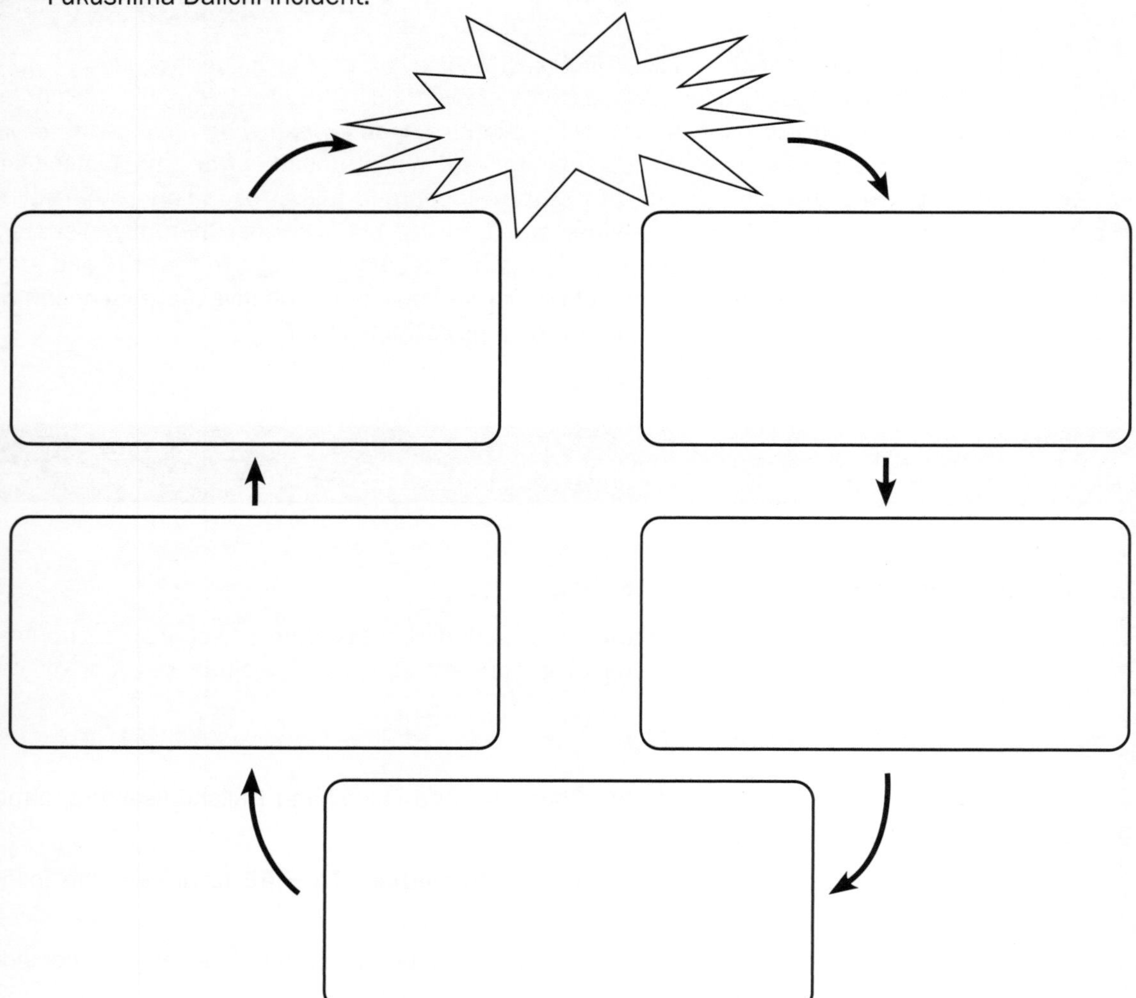

c Account for the different sequence of actions that led to the two disaster events.

ISBN: 9780170425285

Learning Activities

d Give reasons why the two disasters were significant globally.

ISBN: 9780170425285

Q2 Contrasting patterns of radioactive deposition

Learning Activities

Applying a geographic concept: Patterns

Patterns may be spatial (the arrangement of features on the earth's surface) or temporal (how characteristics differ over time in recognisable ways).

Nuclear accidents such as those that occurred at Chernobyl and Fukushima Daiichi are extremely rare. However, when nuclear accidents do occur, radioactive materials (which act in a similar manner to plumes of smoke) can disperse into the atmosphere and travel vast distances. This causes radioactive particles to be deposited on areas affected by the plume, with the volume of radioactive deposition decreasing with distance. When radioactive materials deposit onto land or into the sea, they are absorbed by crops, livestock and marine organisms.

Refer to the definition of patterns above and **Resources 4F** and **4G** to answer this question.

a State the latitude and longitude of Chernobyl to the nearest degree and minute.

b Estimate the area that received more than 3200nSv/h as a result of the Fukushima Daiichi incident.

c Describe in detail one spatial pattern associated with radioactive deposition from the Chernobyl and Fukushima Daiichi nuclear disasters. Tip: use PQE.

i Chernobyl:

ii Fukushima Daiichi:

ISBN: 9780170425285

Learning Activities

d Compare and contrast the magnitude (size and extent) of the Chernobyl and Fukushima Daiichi nuclear disasters.

ISBN: 9780170425285

Q3 Impact of nuclear disasters on the environment

Learning Activities

Applying a geographic concept: Environments

Environments may be natural and/or cultural. They have particular characteristics and features, which can be the result of natural and/or cultural processes. The particular characteristics of an environment may be similar to and/or different from another.

Radioactive materials released from a source such as a nuclear accident can affect the human body through two exposure pathways:

- plume exposure pathway – inhalation of radionuclides, direct irradiation from airborne or deposited radioactive materials
- ingestion pathway – intake of contaminated water or ingestion of contaminated food.

Refer to the definition of environments above and all resources at the end of the chapter to answer the following question.

a Evaluate the extent to which nuclear disasters impact the natural and cultural environment. Support your answer using specific information from **Resource 4H** and the resources at the end of the chapter.

i Impact on the natural environment:

ISBN: 9780170425285

Learning Activities

ISBN: 9780170425285

Learning Activities

ii Impact on the cultural environment:

ISBN: 9780170425285

Disaster Event

Preparedness

Preparedness measures can take many forms including the construction of shelters, implementation of an emergency communication system, installation of warning devices, creation of back-up life-line services (e.g. power, water, sewerage), and rehearsing evacuation plans.

Response

The response phase of an emergency may commence with search and rescue but in all cases the focus will quickly turn to fulfilling the basic humanitarian needs of the affected population. This assistance may be provided by national or international agencies and organisations.

Prevention

Preventive measures are taken on both the domestic and international levels. These are activities designed to provide permanent protection from disasters.

Recovery

The recovery phase starts after the immediate threat to human life has subsided. The immediate goal of the recovery phase is to bring the affected area back to some degree of normalcy.

Mitigation

Mitigation involves any measures taken to limit the impact of disasters, e.g. actions that change the characteristics of a building or its surroundings, i.e. shelters, window shutters, clearing forest around the house. On personal level mainly takes the form of insurance or simply moving house to a safer area.

Resource 4A Disaster management cycle

Resource 4B An abandoned classroom in the nearby settlement of Pripyat near Chernobyl

Resource 4C Chernobyl Nuclear Power Plant

ISBN: 9780170425285

Backgrounder on Chernobyl Nuclear Power Plant Accident

Background

On April 26, 1986, a sudden surge of power during a reactor systems test destroyed Unit 4 of the nuclear power station at Chernobyl, Ukraine, in the former Soviet Union. The accident and the fire that followed released massive amounts of radioactive material into the environment.

Emergency crews responding to the accident used helicopters to pour sand and boron on the reactor debris. The sand was to stop the fire and additional releases of radioactive material; the boron was to prevent additional nuclear reactions. A few weeks after the accident, the crews completely covered the damaged unit in a temporary concrete structure, called the "sarcophagus," to limit further release of radioactive material. The Soviet government also cut down and buried about a square mile of pine forest near the plant to reduce radioactive contamination at and near the site. Chernobyl's three other reactors were subsequently restarted but all eventually shut down for good, with the last reactor closing in 1999. The Soviet nuclear power authorities presented their initial accident report to an International Atomic Energy Agency meeting in Vienna, Austria, in August 1986.

After the accident, officials closed off the area within 30 kilometers (18 miles) of the plant, except for persons with official business at the plant and those people evaluating and dealing with the consequences of the accident and operating the undamaged reactors. The Soviet (and later on, Russian) government evacuated about 115,000 people from the most heavily contaminated areas in 1986, and another 220,000 people in subsequent years. (Source: UNSCEAR 2008, pg. 53)

Health Effects from the Accident

The Chernobyl accident's severe radiation effects killed 28 of the site's 600 workers in the first four months after the event. Another 106 workers received high enough doses to cause acute radiation sickness. Two workers died within hours of the reactor explosion from non-radiological causes. Another 200,000 cleanup workers in 1986 and 1987 received doses of between 1 and 100 rem (The average annual radiation dose for a U.S. citizen is about .6 rem). Chernobyl cleanup activities eventually required about 600,000 workers, although only a small fraction of these workers were exposed to elevated levels of radiation. Government agencies continue to monitor cleanup and recovery workers' health. (UNSCEAR 2008, pg. 47, 58, 107, and 119)

The Chernobyl accident contaminated wide areas of Belarus, the Russian Federation, and Ukraine inhabited by millions of residents. Agencies such as the World Health Organization have been concerned about radiation exposure to people evacuated from these areas. The majority of the five million residents living in contaminated areas, however, received very small radiation doses comparable to natural background levels (0.1 rem per year). (UNSCEAR 2008, pg. 124-25) Today the available evidence does not strongly connect the accident to radiation-induced increases of leukemia or solid cancer, other than thyroid cancer. Many children and adolescents in the area in 1986 drank milk contaminated with radioactive iodine, which delivered substantial doses to their thyroid glands. To date, about 6,000 thyroid cancer cases have been detected among these children. Ninety-nine percent of these children were successfully treated; 15 children and adolescents in the three countries died from thyroid cancer by 2005. The available evidence does not show any effect on the number of adverse pregnancy outcomes, delivery complications, stillbirths or overall health of children among the families living in the most contaminated areas. (UNSCEAR 2008, pg. 65)

Experts conclude some cancer deaths may eventually be attributed to Chernobyl over the lifetime of the emergency workers, evacuees and residents living in the most contaminated areas. These health effects are far lower than initial speculations of tens of thousands of radiation-related deaths.

US Reactors and NRC's Response

The NRC continues to conclude that many factors protect U.S. reactors against the combination of lapses that led to the accident at Chernobyl. Differences in plant design, broader safe shutdown capabilities and strong structures to hold in radioactive materials all help ensure U.S. reactors can keep the public safe. When the NRC reviews new information it takes into account possible major accidents; these reviews consider whether safety requirements should be enhanced to ensure ongoing protection of the public and the environment.

The NRC's post-Chernobyl assessment emphasized the importance of several

Resource 4D Chernobyl nuclear disaster

ISBN: 9780170425285

concepts, including: designing reactor systems properly on the drawing board and implementing them correctly during construction and maintenance; maintaining proper procedures and controls for normal operations and emergencies; having competent and motivated plant management and operating staff; and ensuring the availability of backup safety systems to deal with potential accidents.

The post-Chernobyl assessment also examined whether changes were needed to NRC regulations or guidance on accidents involving control of the chain reaction, accidents when the reactor is at low or zero power, operator training, and emergency planning.

The NRC's Chernobyl response included three major phases: (1) determining the facts of the accident, (2) assessing the accident's implications for regulating U.S. commercial nuclear power plants, and (3) conducting longer-term studies suggested by the assessment.

The NRC coordinated the fact-finding phase with other U.S. government agencies and some private groups. The NRC published the results of this work in January 1987 as NUREG-1250.

The NRC published the second phase's results in April 1989 as NUREG-1251, "Implications of the Accident at Chernobyl for Safety Regulation of Commercial Nuclear Power Plants in the United States." The agency concluded that the lessons learned from Chernobyl fell short of requiring immediate changes in the NRC's regulations.

The NRC published its Chernobyl follow-up studies for U.S. reactors in June 1992 as NUREG-1422. While that report closed out the immediate Chernobyl follow-up research program, some topics continue to receive attention through the NRC's normal activities. For example, the NRC continues to examine Chernobyl's aftermath for lessons on decontaminating structures and land, as well as how people are returned to formerly contaminated areas. The NRC considers the Chernobyl experience a valuable piece of information for considering reactor safety issues in the future.

Discussion

The Chernobyl reactors, called RBMKs, were high-powered reactors that used graphite to help maintain the chain reaction and cooled the reactor cores with water. When the accident occurred the Soviet Union was using 17 RBMKs and Lithuania was using two. Since the accident, the other three Chernobyl reactors, an additional Russian RMBK and both Lithuanian RBMKs have permanently shut down. Chernobyl's Unit 2 was shut down in 1991 after a serious turbine building fire; Unit 1 was closed in November 1996; and Unit 3 was closed in December 1999, as promised by Ukrainian President Leonid Kuchma. In Lithuania, Ignalina Unit 1 was shut down in December 2004 and Unit 2 in 2009 as a condition of the country joining the European Union.

Closing Chernobyl's reactors required a combined effort from the world's seven largest economies (the G-7), the European Commission and Ukraine. This effort supported such things as short-term safety upgrades at Chernobyl Unit 3, decommissioning the entire Chernobyl site, developing ways to address shutdown impacts on workers and their families, and identifying investments needed to meet Ukraine's future electrical power needs.

On the accident's 10th anniversary, the Ukraine formally established the Chernobyl Center for Nuclear Safety, Radioactive Waste and Radio-ecology in the town of Slavutych. The center provides technical support to Ukraine's nuclear power industry, the academic community and nuclear regulators.

The Soviet authorities started the concrete sarcophagus to cover the destroyed Chernobyl reactor in May 1986 and completed the extremely challenging job six months later. Officials considered the sarcophagus a temporary fix to filter radiation out of the gases from the destroyed reactor before the gas was released to the environment. After several years, experts became concerned that the high radiation levels could affect the stability of the sarcophagus.

In 1997, the G-7, the European Commission and Ukraine agreed to jointly fund the Chernobyl Shelter Implementation Plan to help Ukraine transform the existing sarcophagus into a stable and environmentally safe system. The European Bank for Reconstruction and Development manages funding for the plan, which will protect workers, the nearby population and the environment for decades from the very large amounts of radioactive material still in the sarcophagus. The existing sarcophagus was stabilized before work began in late 2006 to replace it with a new safe shelter. The new confinement design includes an arch-shaped steel structure, which will slide across the existing sarcophagus via rails. This new structure is designed to last at least 100 years.

ISBN: 9780170425285

Fact Sheet on Summary of Japan Events in March 2011 and NRC Response

The Nuclear Emergency at Fukushima Daiichi

On Friday, March 11, 2011, a 9.0-magnitude earthquake struck Japan about 231 miles (372 kilometers) northeast of Tokyo off the coast of Honshu Island. The earthquake led to the automatic shutdown of 11 reactors at four sites (Onagawa, Fukushima Daiichi, Fukushima Dai-ni and Tokai) along the northeast coast. Diesel generators provided power until about 40 minutes later, when a tsunami, estimated to have exceeded 45 feet (14 meters) in height, appeared to have caused the loss of all power to the six Fukushima Daiichi reactors. These six reactors have received the majority of the Nuclear Regulatory Commission's attention.

Three Fukushima Daiichi reactors (Units 1-3) were in operation at the time of the earthquake and three (Units 4-6) were shut down for routine refueling and maintenance. As a result of the earthquake, the three operating units automatically shut down as designed. Emergency diesel generators started at all six units, providing power to critical cooling systems. The first large tsunami wave, however, inundated the site and was followed by multiple additional waves, resulting in extensive damage to site facilities including the diesel generators.

Only Unit 6 retained one functional diesel generator, which was used to keep both Units 5 and 6 in a safe, cooled shutdown condition. However, due to lack of diesel generators and offsite power to pump water into Units 1 through 4 to cool the nuclear fuel, as well as the hydrogen gas explosions inside the units, some of the nuclear fuel melted and led to radiation releases. In the absence of early response of offsite assistance, which appears to have been hampered by the devastation in the area, among other factors, each unit eventually lost its cooling capability. Sources of water were finally brought in to cool the reactors and work continues to stabilize these plants.

NRC in Action

On Friday morning, shortly after the earthquake and tsunami occurred, the NRC started monitoring the situation. Later that afternoon, the NRC activated and staffed its Emergency Operations Center at headquarters in Rockville, Maryland, to closely monitor the Japan events and assess the potential impact on U.S. nuclear plants and materials, particularly those on the West Coast, and in Hawaii, Alaska, and U.S. Territories in the Pacific. The agency began interactions with its Japanese regulatory counterparts and dispatched two experts to Japan to help at the U.S. Embassy in Tokyo. By Monday, March 14, the agency had dispatched a total of 11 NRC staff to provide technical support to the American Embassy and the Japanese government. The NRC was there to assist the Japanese government and respond to requests from its Japanese regulatory counterparts. The agency also provided support to the U.S. ambassador and the U.S. government assistance effort.

On Wednesday, March 16, the NRC collaborated with other U.S. government agencies and through the U.S. Ambassador in Japan advised American citizens to evacuate within a 50-mile range around the Fukushima plant. The 50-mile evacuation recommendation was made in the interest of protecting the health and safety of U.S. citizens in Japan based on the information available at that time. The agency alerted its licensees to the events in Japan and sought to assist them in considering the ramifications of a similar event for their facilities and to take site-specific actions, as appropriate. The NRC issued instructions to its inspectors for immediate independent assessments of each plant's level of preparedness. The instructions covered extensive damage mitigation guidelines, station blackout, and seismic and flooding issues, as well as severe accident management guidelines.

Post Event Activities

Since the events at Fukushima began to unfold in early March, the NRC has been working to understand the events in Japan anad relay important information to the U.S. nuclear power plants. Not long after the emergency began, the NRC established a task force of senior NRC experts to determine lessons learned from the accident and to initiate a review of NRC regulations to determine if additional measures needed to be taken immediately to ensure the safety of nuclear power plants in the United States. The task force issued its report on July 12, 2011, which concluded that there was no imminent risk from continued operation and licensing activities. The Task Force also concluded that enhancements to safety and emergency preparedness are warranted and made a dozen recommendations for Commission consideration. The Commission is currently considering both short-term and longer-term actions to ensure nuclear plant safety in the United States.

For more information on NRC's actions, visit the NRC website and the Japan Nuclear Accident – Implementing Lessons Learned from Fukushima Web page.

September 2011

Resource 4E Fukushima Daiichi nuclear disaster

ISBN: 9780170425285

Surface ground deposition of caesium-137 released in Europe after the Chernobyl accident

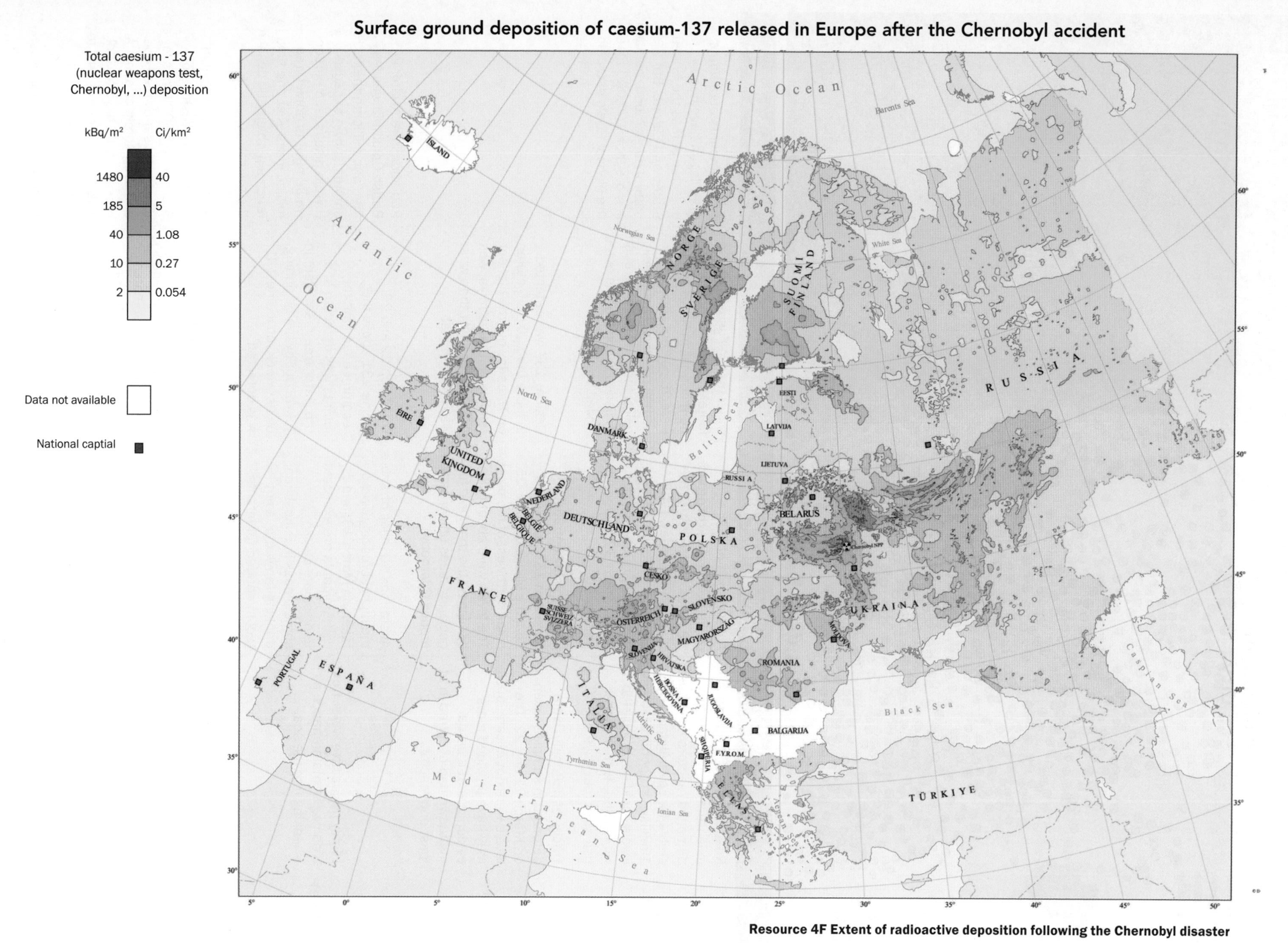

Resource 4F Extent of radioactive deposition following the Chernobyl disaster

ISBN: 9780170425285

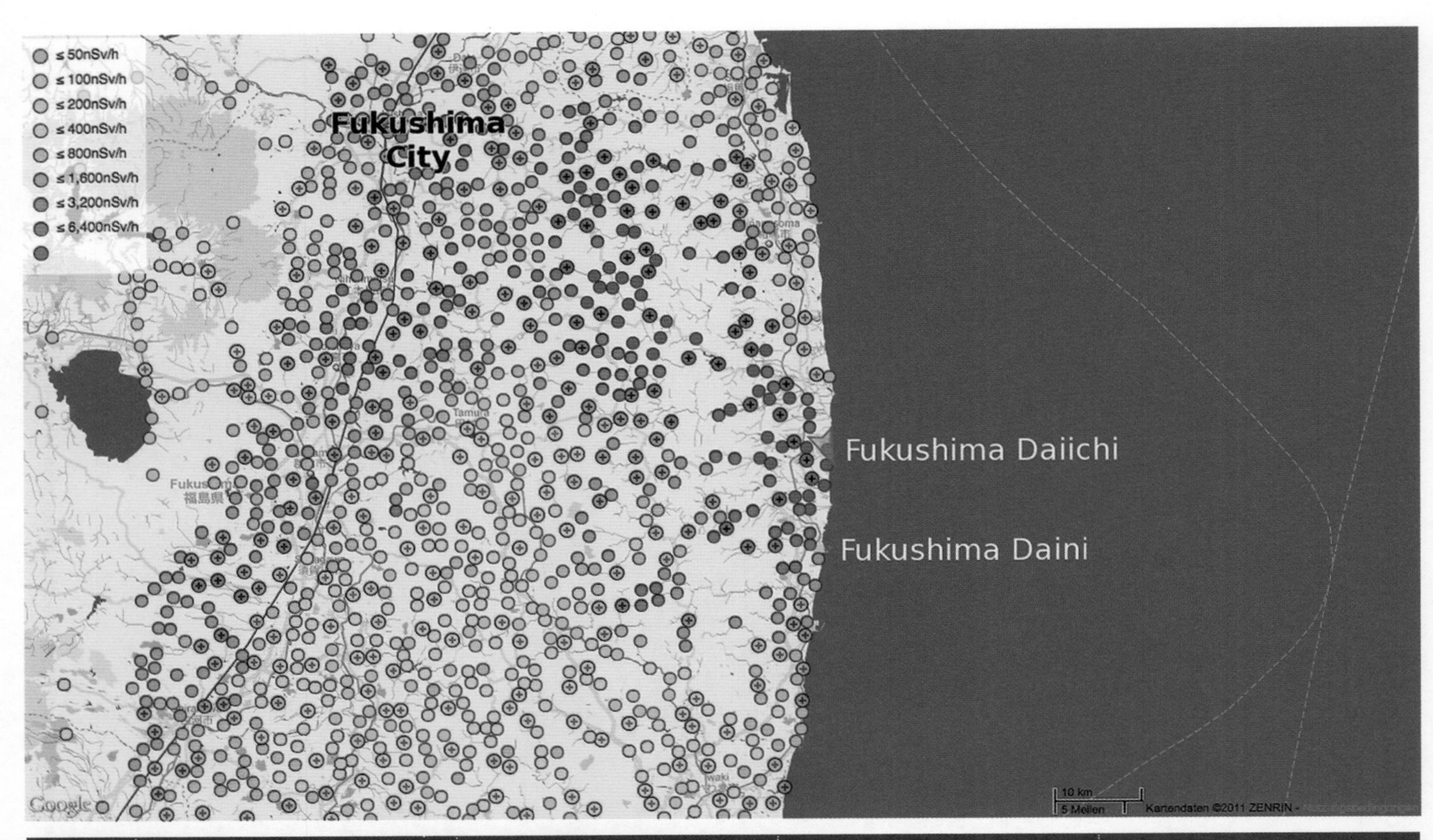

	At the Plant	Near the Plant	In Japan	Around the World
Air	Readings about a quarter mile from the most heavily damaged reactors have been stable for several days. Near 1 millisievert per hour, these levels could be associated with slightly higher cancer risk after four days.	A daily dose of 0.8 millisieverts was recorded 19 miles northwest of the plant on Thursday. I.A.E.A. guidelines recommend temporary relocation if levels reach 30 millisieverts per month.	Other than in Fukushima and Ibaraki, levels are not far from normal. In Tokyo, levels were 25 percent above the normal range on Friday, well below the level of background radiation in some areas of the United States.	Trace amounts of radiation from Japan have been detected across the United States and Europe. But natural background radiation is more than 100,000 times the highest levels detected.
Soil	Traces of plutonium were detected in samples taken on March 21 and 22. The levels were not unsafe, but may have provided more evidence that a partial meltdown had occurred in at least one reactor.	Very high concentrations of cesium 137 were found near the village of Iitate, 25 miles northwest of the plant. The levels were about twice as high as the threshold for declaring areas uninhabitable around Chernobyl.	Cesium 137 was detected in more than 10 prefectures on Thursday but the highest reading, in Utsunomiya, was 4000 times lower than what was found in Iitate.	
Water	Highly radioactive water from a damaged maintenance pit is leaking into the sea.	At stations 19 miles offshore, the highest readings were taken on March 13, and contaminants are expected to dissipate quickly. At some places in Fukushima, drinking tap water is not recommended for infants.	On March 22 and 23, iodine 131 above the recommended limit for infants was detected at a tap water treatment plant in Tokyo. But by the beginning of last week, no iodine 131 was detected.	Radiation in rainwater in British Columbia was less than one millionth the amount shown to cause thyroid diseases. A person would have to drink three million glasses at one time to reach a problematic dose in the thyroid.
Food	Fishing has been banned in the evacuation zone.	Radioactive cesium was detected in broccoli in Fukushima Prefecture well above the country's limit. The estimated increase in cancer risk of eating two unwashed pounds is about two chances in a million.	Radioactive cesium was detected in beef from Tenei at a level just above Japan's legal limit. The estimated increase in cancer risk of eating two pounds is about one chance in 10 million.	Radiation levels detected in milk from Washington State were 5000 times lower than limits set by the Food and Drug Administration. A person would have to drink 1552 gallons of this milk to reach the limit.

Resource 4G Extent of radioactive deposition following the Fukushima Daiichi disaster

ISBN: 9780170425285

Effects of radiation on the human body

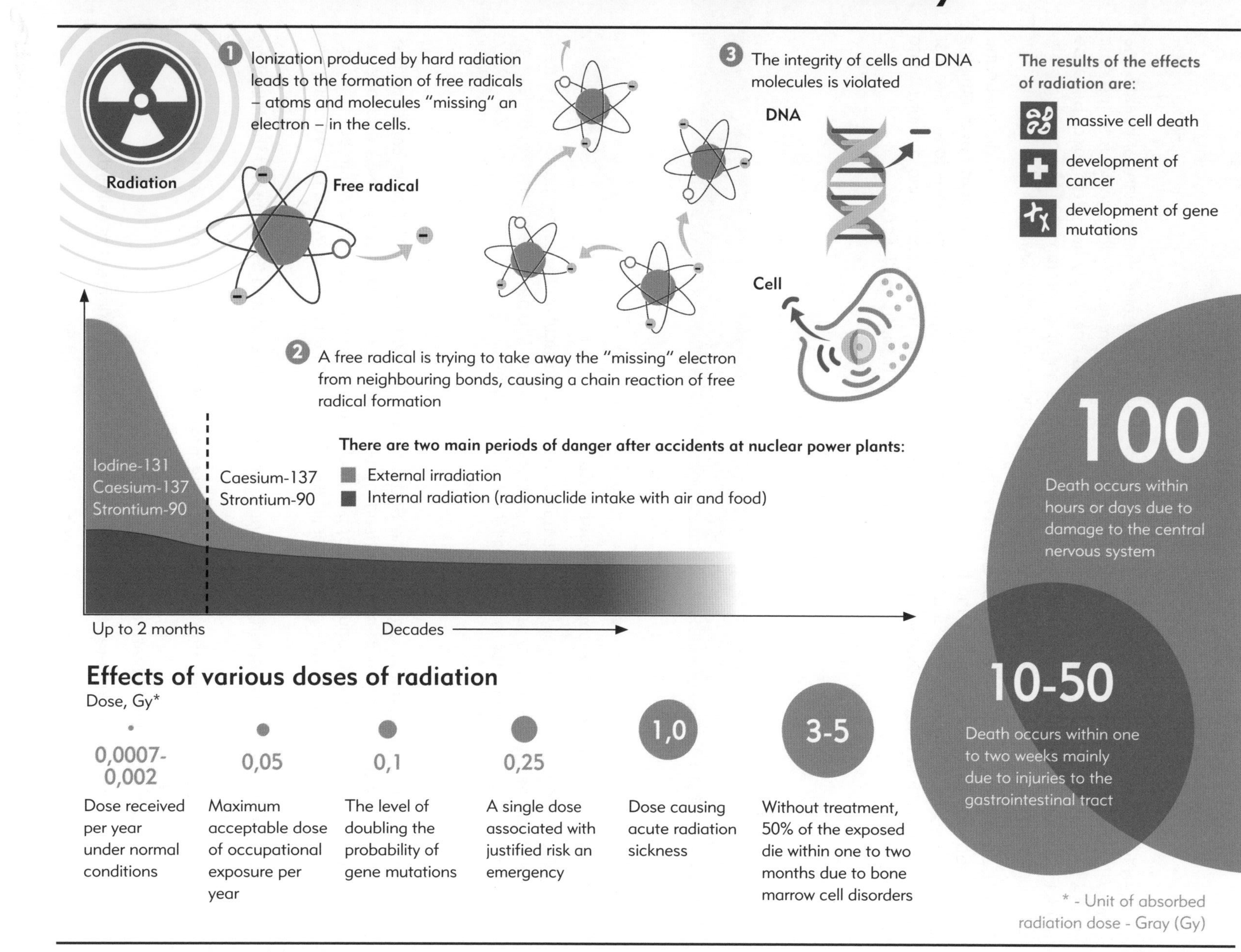

Resource 4H Effects of radiation on the human body

ISBN: 9780170425285

Destination Fiordland

Located in Fiordland National Park, Milford Sound is by virtue of its isolation and natural beauty one of New Zealand's most visited tourist attractions. Separated from more populous regions to the east by the Southern Alps, overland access to the Sound is limited to one windy mountain road. Because of its location in Fiordland National Park, which prevents substantial accommodation from being built, approximately 90 percent of its visitors are day trippers visiting from Queenstown, a 12-hour round trip.

These issues have resulted in a number of serious proposals on how to better connect Milford Sound to neighbouring urban settlements and how to increase tourism in the region without compromising the sustainability of the area. The Fiordland Link Experience is one proposal that aims to achieve just this by combining several innovative transport options (catamaran, terrain coach and monorail) into one trip. However, the proposal (like those that came before) is a controversial one.

Q1 Natural environment of Fiordland

Learning Activities

Applying a geographic concept: Environments

Environments may be natural and/or cultural. They have particular characteristics and features, which can be the result of natural and/or cultural processes. The particular characteristics of an environment may be similar to and/or different from another.

Fiordland is a geographic region of the South Island (Te Waipounamu) of New Zealand. The topography of the region is characterised by temperate rainforests, deep lakes and ocean-flooded valleys, waterfalls and snow-capped mountains. Fiordland contains a number of fiords (long, narrow sea inlets with steep sides, created by glacial erosion), of which Milford Sound is its most famous.

Refer to the definition of environments above and **Resources 5A** and **5B** to answer the following questions.

a Complete the précis map on the following page by accurately locating and labelling the following features:

- **i** two recreational features
- **ii** three natural features that hinder accessibility to the settlement of Milford Sound
- **iii** three cultural features that aid accessibility to Milford Sound.

ISBN: 9780170425285

Learning Activities

Key

ISBN: 9780170425285

Learning Activities

b Circle the approximate representative fraction of the précis map.

1:250,000 1:50,000 1:25,000 1:5000

c In the space below, construct a detailed cross-section between six-figure grid reference 950360 and 030360.

d The natural environment of the Fiordland region can be perceived as both an advantage and a disadvantage for its potential as a tourist destination. Give examples of each using evidence from the resources.

i One advantage of Fiordland as a tourist destination:

ii One disadvantage Fiordland as a tourist destination:

ISBN: 9780170425285

Q2 Fiordland Link Experience

Learning Activities

Applying a geographic concept: Perspectives

Perspectives are ways of seeing the world that help explain differences in decisions about, responses to, and interactions with environments. Perspectives are bodies of thought, theories or worldviews that shape people's values and have built up over time. They involve people's perceptions (how they view and interpret environments) and viewpoints (what they think) about geographic issues. Perceptions and viewpoints are influenced by people's values (deeply held beliefs about what is important or desirable).

In 2011, the Department of Conservation notified its intention to grant Riverstone Holdings Ltd a 49-year concession to build a $200 million three-stage transport link between Queenstown and Te Anau Downs.

Fittingly called the Fiordland Link Experience, the project includes a controversial 43-kilometre monorail stage that would run through the Snowdon Forest Conservation Area, a place of high ecological value that is administered by the Department of Conservation. Snowdon Forest includes red tussock grasslands, which are of international importance, and is recognised in the World Heritage status of the site. Other vegetation types along the proposed monorail route, such as tall red beech forest, are of national importance as habitats for kaka, bats and other native species. The forest also supports a variety of threatened bird species.

Refer to the definition of perspectives above and **Resources 5C – 5F** to answer this question.

a Choose two groups of people/groups with different perspectives about the monorail proposal.

i Show their viewpoints on the continuum below.

Supportive of the monorail proposal ⟷ Against the monorail proposal

ii Justify why you have placed these two groups where you have in **i**. Justify your answer with specific detail.

ISBN: 9780170425285

Learning Activities

ISBN: 9780170425285

Q3 Sustainable tourism

Applying a geographic concept: Sustainability

Sustainability involves adopting ways of thinking and behaving that allow individuals, groups and societies to meet their needs and aspirations without preventing future generations from meeting theirs. Sustainable interaction with the environment may be achieved by preventing, limiting, minimising or correcting environmental damage to water, air and soil, as well as considering ecosystems and problems related to waste, noise, and visual pollution.

The Resource Management Act (RMA) is the main piece of legislation that sets out how we as New Zealanders should manage our environment. The Act is based on the idea of the sustainable management of resources by encouraging land users to mitigate the effects of their activities and consider the long-term future of our environment. Before a business can undertake a development involving the environment, it first must prove that it can manage the effects of the development and ensure the environment will not suffer.

a With reference to the definition of sustainability above and **Resources 5E – 5H**, comprehensively analyse the environmental sustainability of Riverstone Holdings' monorail proposal.

ISBN: 9780170425285

Learning Activities

ISBN: 9780170425285

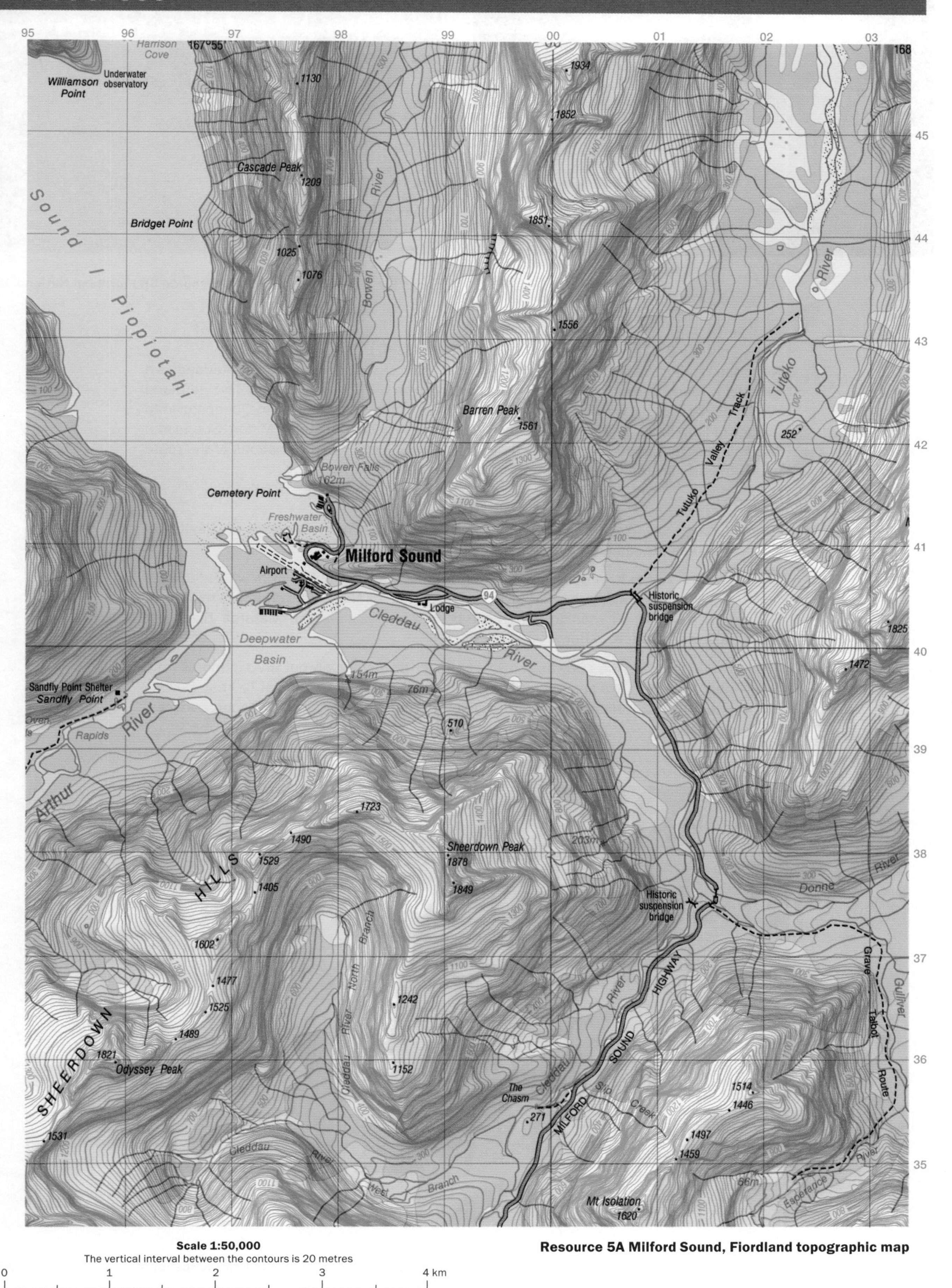

Resource 5A Milford Sound, Fiordland topographic map

ISBN: 9780170425285

Resource 5B Fiordland National Park

Riverstone Holdings Limited proposes the 'Fiordland Link Experience', which is intended as a high-quality tourism experience connecting Queenstown with Te Anau Downs. The Fiordland Link Experience involves a surface transport system in three sections from Queenstown to the edge of Lake Te Anau. The proposal would improve access to Te Anau, Milford Sound and Fiordland generally by reducing travel times from Queenstown. These sections of the route are on land administered by the Department of Conservation.

The three sections are:

- Across Lake Wakatipu: A 20 km trip by catamaran from Queenstown Bay to Mt Nicholas Station (at White's Bay).
- Mt Nicholas to Kiwi Burn: A 45 km trip by an all-terrain vehicle from wharf facilities at White's Bay along the Von River valley (on the Mt Nicholas/Von Road) to Mavora Lakes Road and then to a purpose-built terminus near the Mararoa River just upstream from Kiwi Burn.
- Kiwi Burn to Te Anau Downs: A 43.8 km trip by an electrically powered monorail from near Kiwi Burn to a second terminus at the edge of Lake Te Anau at Te Anau Downs. From there travellers could connect to other road- or lake-based transport services.

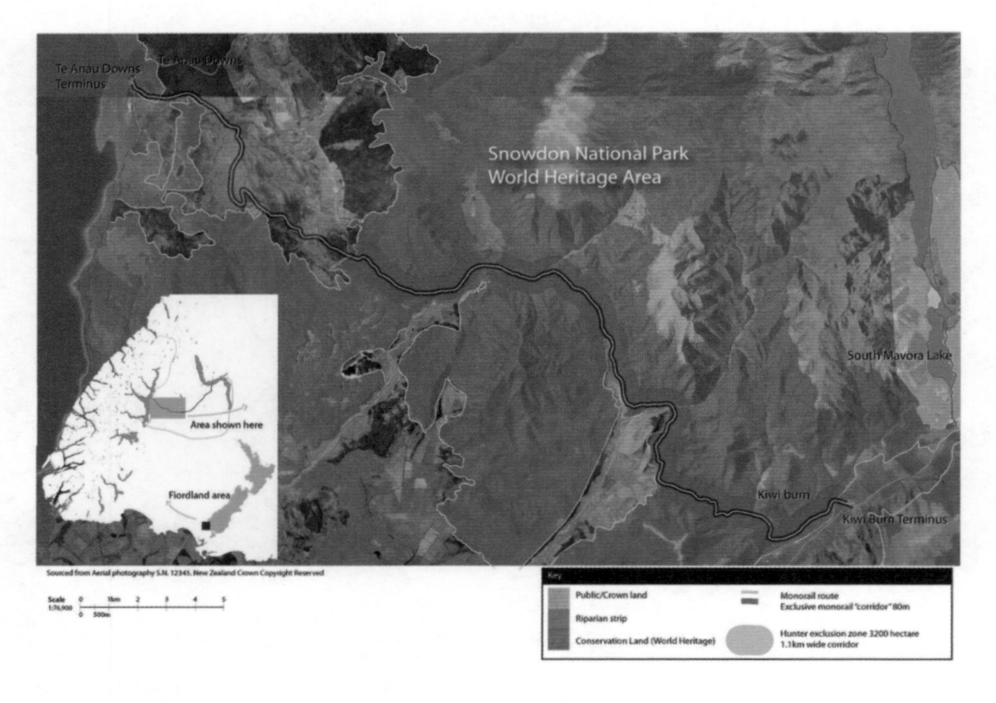

Resource 5C Fiordland Link Experience route

ISBN: 9780170425285

Otago Daily Times

Riverstone will 'sell' Te Anau

The proponent of the proposed $250 millon monorail project linking Queenstown and Te Anau is vowing the company will become Te Anau's biggest promoter.

Riverstone Holdings chief executive Bob Robertson, of Wanaka, believed the monorail, 16 years in the planning, would lift local tourism and the area's economy.

"I think Te Anau locals will probably think it might be negative for them because people are currently travelling through Te Anau."

One of the biggest criticisms of the Milford Dart Tunnel proposed last month was Te Anau would miss out on potential visitors, but Mr Robertson said his company would spend about $5 million a year promoting Te Anau and the "new visitors" would be stopping there for more than "just a coffee".

"We will spend more than anyone has before on selling the destination. We will be an investor locally. We would become the biggest sponsor."

He said the company had put more than $1 million into Wanaka's economy and he would like to do the same for Te Anau.

The monorail was part of Riverstone Holdings' visitor transport package, The Fiordland Link Experience, which would include a catamaran across Lake Wakatipu, an all-terrain ride on back-country roads and the 80km/h monorail trip.

The company had a project budget of $175 million to $200 million, with another $50 million for "long-term upgrades" such as transport facilities running into Te Anau.

If consent was granted, he expected a completion timeframe of within two years.

Mr Robertson hoped to establish a fund to help finance the project, with four domestic investors already expressing an interest in contributing.

"I know if I've got the right economic model... funding is not an issue."

It was not about shortening the trip from Queenstown to Te Anau, he said.

"We're not in a hurry. We're not promoting hurry. We are promoting [an] experience while still delivering a benefit of speed.

"We will get them there quicker and they will stay longer."

The initial idea came from Malaysia, which was where most of the rolling stock and electronics would be built, he said.

However, Te Anau's gain could be Kingston's loss, as travellers choosing the monorail would bypass State Highway 6.

Kingston Community Association chairwoman Annetta Dalziel said local businesses had already shown some concern.

"To me, it seems a terrible way [to travel]. Just whipping people here and there and not giving them the time to meet New Zealanders.

Te Anau Community Board chairman Alistair Jukes, opposed to the Milford Dart Tunnel, said yesterday the monorail was "not as bad".

"It's certainly got its merits. I'm not really in opposition to it."

Southland District Council mayor Frana Cardno said both the monorail and Milford Dart Tunnel were "all about getting to Milford faster".

"New Zealand tourism talks about quality tourism. Is this what we call quality tourism? "

The Department of Conservation had yesterday received 61 submissions for the the Milford Dart Tunnel with submissions closing in January.

Mr Jukes said the Te Anau Community Board would be submitting against the Milford Dart Tunnel.

Public submissions on the monorail will close in February.

Resource 5D Riverstone will 'sell' Te Anau

ISBN: 9780170425285

Impact identified by Forest & Bird

The monorail is proposed to run from the Mararoa River, just below the Mavora Lakes in Southland, through Snowdon Forest in Te Wahipounamu – South West New Zealand World Heritage Area out to Lake Te Anau. It involves clear-felling old growth beech forest and disturbing rare red tussock grasslands to construct two parallel roads. The public and tourists already have access by public roads to these areas in the National Parks and the jury is out as to whether they would alleviate, or even add to, congestion at Milford.

Potential adverse effects include:

- Considerable earthworks to construct monorail and access tracks – to be converted to a bike trail following construction.
- Clearance and edge effects of at least 68 ha of wildlife-rich forest, approximately 19,555 trees, including around 76 very large red beech trees in significant lowland red beech forest.
- Significant adverse effects on rare unmodified low-altitude red tussock valley grasslands – about 4.5 ha.
- Significant adverse effects on threatened species including the nationally endangered long-tailed bat, threatened mohua, and kaka, and other forest birds. (Around 5500 large trees would be cleared.) As bats roost in clusters of trees, DOC says if such a cluster were felled the effects on the bat population could be catastrophic.
- Impossible to accurately assess the impacts due to use of an envelope approach over a 200 m wide corridor rather than precise routes.
- Increased weeds and pests.
- Sedimentation pollution runoff to affected small streams and rivers including the Mararoa and Kiwi Burn Rivers.
- Potentially significant landscape effects, described by the technical landscape audit as significantly compromising the outstanding natural values of the landscapes of the Snowdon Forest part of Te Wahipounamu – South West New Zealand World Heritage Area.
- Loss of the popular family tramp to the Kiwi Burn hut (hut is to be relocated).

Resource 5E Impact identified by Forest & Bird

ISBN: 9780170425285

The case for derailing the plan

OPINION: Conservation Minister Nick Smith should turn down plans for a monorail through Fiordland forest, writes Bill Jarvie.

Plans for a scheme that would see a monorail pushed through remote Fiordland beech forest and river valleys simply do not stack up, and would be a disaster for the region and the country.

The concept of a three-stage boat-bus-monorail link from Queenstown to the road to Milford Sound is not, as Bob Robertson claims, an initiative conceived to address the current needs of tourism (Scaremongers putting forest monorail at risk, Sept 12).

It was first promoted more than 17 years ago. In its present format it has been around for nearly a decade.

The "experience" would not reduce the travel time for Milford Sound tourists. Once off-loaded from the monorail they would be bussed for another 1 1/2 hours to Milford. In one day they would endure a minimum of 12 changes of transport in a convoluted return trip.

What has changed from inception is that the intended destination of the monorail is to the company's hotel/restaurant site at Te Anau Downs, avoiding tourist-dependent Te Anau.

Te Anau is vibrant and superbly set up with international class hotels, award-winning motels and restaurants. It is the most appropriate destination en route to Milford Sound.

Again, none of the company's publicity mentions the existing tourism and accommodation options and the several towns that would be rendered backwaters should this proposal capture the tourist numbers it would require to be economic.

In order to achieve the numbers, Riverstone would construct more than 29 kilometres of elevated concrete and steel monorail plus permanent parallel construction-maintenance roading through remote World Heritage forest and river valleys.

Contrary to the images that Riverstone has distributed, the entire monorail proposal is one of industrial scale development with massive structures only seen in large cityscapes.

Anyone who is familiar with southern beech forests knows that a six-metre-wide clearance for a high-speed train is nonsense.

A minimum of 20,000 trees would be felled, including red beech several hundred years old. Forest cleared in continuous swaths of undisclosed width would be cut across hillsides in what the Department of Conservation describes as an "outstanding natural landscape", with ongoing felling of potentially dangerous trees and the regular cutting of forest regrowth.

The river flats would be scarred with the continuous elevated beams and supporting piers every 20 metres.

The adjacent gravel road would carry construction vehicles such as trucks, excavators and drilling rigs. It would be permanent because of the need for maintenance and emergency evacuation of tourists tens of kilometres from any public road.

Mr Robertson's comparison with the Cairns Skytrail is amusing. The Skytrail is a leisurely traverse of the tree tops through what was already a developed landscape. Trees were specifically avoided, not felled.

Passengers can step out at mid-stations to experience the forest interior from boardwalks and lookouts, and spend time in an interpretation centre.

To meet its timetable the monorail ride would be at speeds up to 90km/h through a blur of forest interior.

Mr Robertson attempts to dress this up as an environmental experience for tourists seeking the 100% pure New Zealand. *The Dominion Post* has aptly described the ride: "its whizzing caravan of gawkers an insult to the thing that brought them in the first place." (Editorial, June 22)

The Snowdon Forest Conservation Area is already a highly valued and accessible introduction to New Zealand's natural environment.

DOC publicity describes the monorail's starting point as "one of the best opportunities in Southland for introducing families to tramping and the outdoors".

The very area the monorail would cut through is where people who actually value it walk in, and pass through on foot if heading to more demanding country. The proposed development would destroy the very heart of what makes it so special to those who put some personal effort into gaining the experience.

There are already better means for tourists of all capabilities to experience what sets us apart from the rest of the world.

There are hundreds of non-destructive concessionaires, many of whom have been vocal in their opposition to this proposal.

To improve the value of the tourism sector we need to preserve our unique environment in order to attract higher value longer-stay visitors rather than go for volume by appealing to mass market short-term visitors.

New Zealand and the world have limited resources so our approach should be to protect what we have rather than continuously erode it away around the edges.

Mr Robertson should be concerned about Conservation Minister Nick Smith's imminent decision on the concession application. DOC commissioned independent audits on the proposal's engineering credibility and environmental impacts have repeatedly warned of significant and unresolved concerns over logistical difficulties and unmitigated impacts on this untouched part of the World Heritage Area.

Bill Jarvie is the chairman of Save Fiordland.

Resource 5F The case for derailing the plan

ISBN: 9780170425285

Conservation management strategy

The adaptive management approach taken to the Fiordland monorail – designed to ensure good decision-making throughout the life of the project – will be bound by appropriate conditions guiding the development and enforcement of a range of management plans.

The management plans guiding the project will include:

» **Implementation Protocol – outlining the relationship between Riverstone and DOC**

» **Construction Management Plan – guide construction to minimise environmental impact**

» **Recreation Users Management Plan – to ensure effects on other users are avoided, minimised or mitigated**

» **Vegetation and Habitat Management Plan – to manage the effects on terrestrial ecology values during both the construction and on-going operation of the monorail and associated facilities**

» **Operational Management Plan – to manage the on-going use and maintenance of the monorail and associated facilities.**

In addition, a wide range of monitoring and management programmes will be undertaken to protect the environment. The Environmental Monitoring and Management Plan will:

» **Ensure rehabilitation of the worked area during construction has been successful;**

» **Monitor the effects of the monorail operation on indigenous flora and fauna in the affected area;**

» **Prevent the establishment and spread of weeds;**

» **Prevent increased use of the area by pests;**

» **Ensure that the ecological integrity of the site is preserved as required by the DOC Conservation Management Strategy (2000).**

Resource 5G Conservation management strategy

Resource 5H

Description: Martin Doyle cartoon: Monorail through the heart. Rare native species like the mohua have had 'Threatened' beside their names. Thanks to the proposed monorail through their habitat we can safely upgrade that to 'Assaulted'.

ISBN: 9780170425285

City Rail Link

With a population of 1.4 million (2013) and accounting for 32 percent of the country's total population, Auckland is New Zealand's largest, most populous urban area. It is also a rapidly growing city and its population is expected to reach the 2 million mark by 2031.

Aucklanders are known for their dependence on private cars for day-to-day mobility. Currently, 85 percent of trips in Auckland are made by private car, and around 15,000 extra cars join Auckland's roads every year. Improving public transport options and connections along key transport corridors while encouraging Aucklanders to use public transport will become a priority for local planners as they seek to reduce traffic congestion into the future.

Q1 Inter-census growth and mobility

Learning Activities

Applying a geographic concept: Patterns

Patterns may be spatial (the arrangement of features on the earth's surface) or temporal (how characteristics differ over time in recognisable ways).

Recent census findings indicate that the Auckland region experienced the greatest growth between 2006 and 2013 censuses with a population increase of 110,592 people. To put that in perspective, that is 51.6 percent of all of the growth that has occurred in New Zealand since the last census and is similar to having added the entire city of Dunedin (120,246) or twin cities of Napier-Hastings (125,000) to the region.

Refer to the definition of patterns above and **Resources 6A** and **6B** to answer the following questions.

a Describe in detail one spatial and one temporal pattern associated with Auckland's population growth. Tip: use PQE.

i Spatial pattern:

ISBN: 9780170425285

Learning Activities

ii Temporal pattern:

b Describe in detail one spatial and one temporal pattern associated with the Auckland's population density. Tip: use PQE.

i Spatial pattern:

ii Temporal pattern:

ISBN: 9780170425285

Q2 The scale of Auckland's transport congestion

Learning Activities

Applying a geographic concept: Change

Change involves any alteration to the natural or cultural environment. Change can be spatial and/ or temporal. Change is a normal process in both natural and cultural environments. It occurs at varying rates, at different times and in different places. Some changes are predictable, recurrent or cyclic, while others are unpredictable or erratic. Change can bring about further change.

Population growth and increasing mobility to and from Auckland's CBD has caused transport systems in Auckland to become overburdened and inefficient. Lack of investment in public transport means that most Aucklanders are dependent on private cars as their primary means of transportation. This dependence has, over time, led to heavily congested roads and motorways.

Refer to the information in **Resources 6C – 6E** answer the following questions.

a As the population of Auckland continues to increase, so to does the day-to-day mobility of its population.

i Use an appropriate statistical mapping technique to display the mobility of Auckland's population on the map below. Use appropriate mapping conventions.

ISBN: 9780170425285

Learning Activities

ii With reference to the statistical map you have drawn, explain in detail the pattern of mobility in Auckland. Tip: use PQE.

b Refer to **Resource 6D**. Calculate the percentage increase in vehicle kilometres travelled in the period from 2007/08 to 2016/17.

c Use an appropriate graphing technique to illustrate the information in **Resource 6D**. Tip: use SALTS.

d Account for the trend shown in the graph you constructed in **c**.

ISBN: 9780170425285

Q3 Shaping Auckland's transport future

Learning Activities

Applying a geographic concept: Sustainability

Sustainability involves adopting ways of thinking and behaving that allow individuals, groups and societies to meet their needs and aspirations without preventing future generations from meeting theirs. Sustainable interaction with the environment may be achieved by preventing, limiting, minimising or correcting environmental damage to water, air and soil, as well as considering ecosystems and problems related to waste, noise, and visual pollution.

Projected population growth will only exacerbate congestion unless a more sustainable and integrated transport network that enables people and goods to move efficiently is put in place. One proposal that will potentially alleviate congestion within the Auckland urban area is the City Rail Link (CRL).

Refer to the definition of sustainability above and **Resources 6F – 6J** to answer this question.

a Evaluate how the introduction of the City Rail Link will change transport mobility within and around Auckland City.

ISBN: 9780170425285

Learning Activities

ISBN: 9780170425285

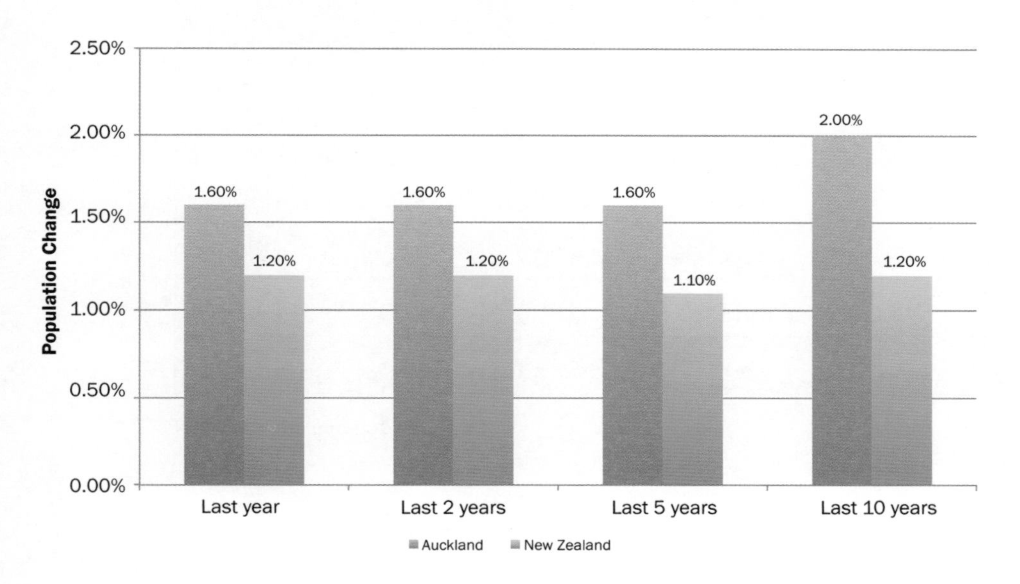

The Auckland region's population is growing and may reach 2 million by 2031. International migration is a bigger contributor to the region's population growth than it is for the rest of New Zealand. However, since the period 1996–2001, more people have moved from the Auckland region to other parts of New Zealand than in the other direction, but this is a new phenomenon.

Resource 6A Population growth (2011)

In the period since the 2013 Census, Auckland's regional population has increased at an average rate of 1.5 percent per year. That is more than the average growth rate of 1.2 percent per year Auckland experienced between the 2006 and 2013 censuses. Accordingly, Auckland's share of the country's growth is expected to account for three-fifths of New Zealand's population growth between 2018 and 2043.

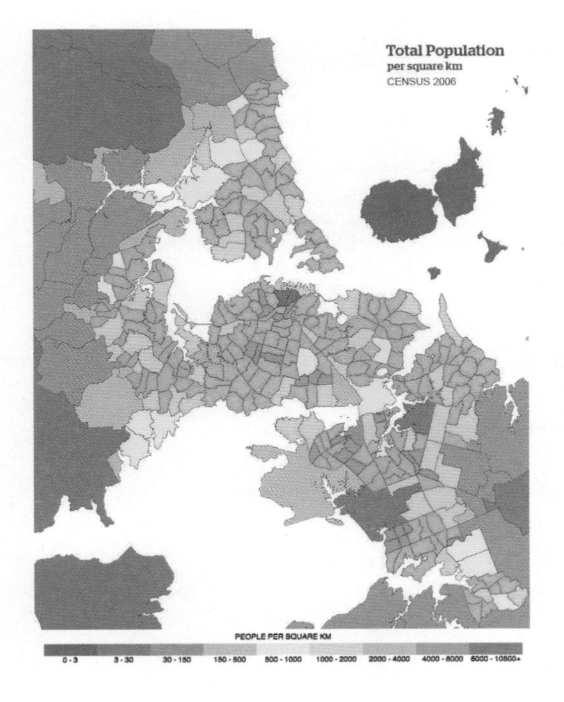

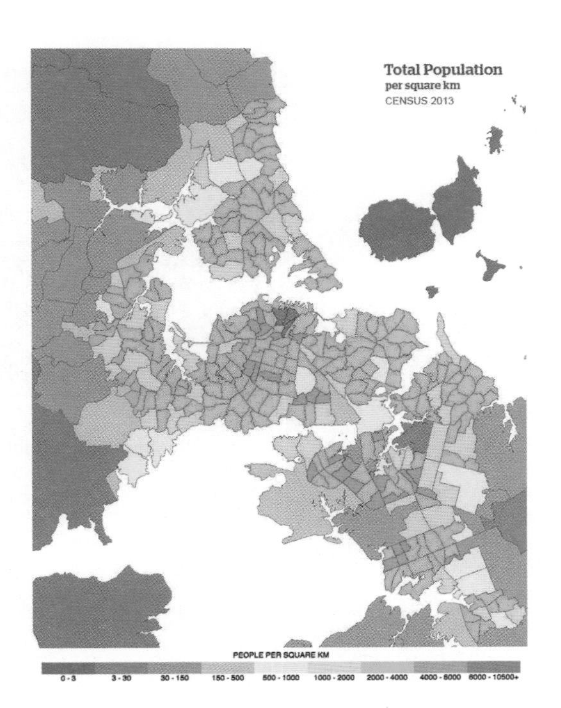

Resource 6B Population density

ISBN: 9780170425285

The information in the table below shows journey to work patterns around Auckland.

Place of residence	Place of work					
	North	West	CBD	Isthmus	South	Total
North	57,000	4,700	13,000	13,400	3,800	91,900
West	4,700	20,700	8,200	17,000	3,500	54,100
CBD	500	200	1,400	1,900	400	4,400
Isthmus	6,500	5,000	28,500	59,700	15,200	114,900
South	2,700	1,200	8,000	30,800	76,500	119,200
Total	71,400	31,800	59,100	122,800	99,400	384,500

Resource 6C Population mobility

The table below shows that vehicle kilometres travelled (VKT) per person on the state highway network in Auckland is increasing at a constant annual rate of approximately 1 percent per year.

Auckland State Highway Vehicle Kilometres Travelled (VKT) per person		
Year	**VKT**	**% Increase**
2007/08	4,147,364,000	
2008/09	4,103,784,000	1.01
2009/10	4,193,135,000	0.98
2010/11	4,282,705,766	0.98
2011/12	4,691,627,488	0.91
2012/13	4,481,953,027	1.05
2013/14	4,600,391,807	0.97
2014/15	4,747,047,256	0.97
2015/16	4,834,622,601	0.98
2016/17	5,028,095,129	0.96

Resource 6D Vehicle kilometres travelled (VKT) per person

ISBN: 9780170425285

Commuter hotspots revealed

A study has revealed the worst suburbs for Auckland commuters and the news is bad for early risers on the North Shore.

According to new figures it takes Takapuna residents almost three times as long to travel to downtown Auckland during peak morning traffic times than outside those hours.

The travel times were revealed in a study by navigation services provider TomTom and showed North Shore isn't friendly to morning commuters with Albany, Mairangi Bay and Devonport the next-worst suburbs in that order.

Between 7am and 9am it takes Takapuna residents 28.25 minutes to get to the Sky Tower - more than 18 minutes longer than it takes between 1am and 5am when traffic is lightest.

Papatoetoe and Pakuranga rounded out the top six which all clocked increased travel times of more than 100 percent.

Ponsonby is the place to live for people who want to get quickly from A to B in the morning, with an increase of only 36 percent travel time in rush hour.

Mt Eden and Parnell were the other two suburbs to come in with less than a 50 percent increase in travel time at 40 percent and 38 percent respectively.

In the evenings, Remuera is the worst suburb with travel times more than doubling from 8.26 minutes to just over 19 between 4pm and 6pm.

It was followed by Pakuranga, Manukau, Onehunga and Papatoetoe which all recorded increases in travel times of more than 100 percent.

Only Devonport and Glendowie recorded 50 percent or less.

To come up with the results, TomTom studied travel statistics from users over a nine-month period from June last year where it clocked travel times for users from various suburbs on weekdays.

The average travel time during peak traffic was compared with average travel time when traffic was free flowing.

Phil Allen, a spokesman for AA-owned GeoSmart which provides real-time information on roads and traffic flows, said the information was aimed at emphasising the congestion on our roads.

"It's about how do you better spread the load on the road," he said.

New roads would not keep up with increased traffic loads and spreading traffic out on different routes would help ease congestion - an issue which costs Auckland around $1 million a year.

In-car navigation systems helped people do this, he said.

Resource 6E Commuter hotspots revealed

ISBN: 9780170425285

The City Rail Link (CRL) will significantly improve the Auckland rail network. It is a proposed 3.5-kilometre underground rail link between Britomart and Mt Eden Station on the western rail line, which will provide three new stations in the central city. It will address the capacity constraints at Britomart, enable future increases in rail service frequency across the whole rail network, and add new rail lines to the network (such as rail to the airport).

Eighty percent of submitters to the Draft Auckland Plan who referred to the CRL supported its construction. The CRL is the foremost transformational project in the next decade. It creates the most significant place-shaping opportunity, as the entire city centre would be within 10 minutes' walk of a railway station. As well, many more rail trips across Auckland could take place as a continuous ride without needing to transfer.

Resource 6F CRL light rail

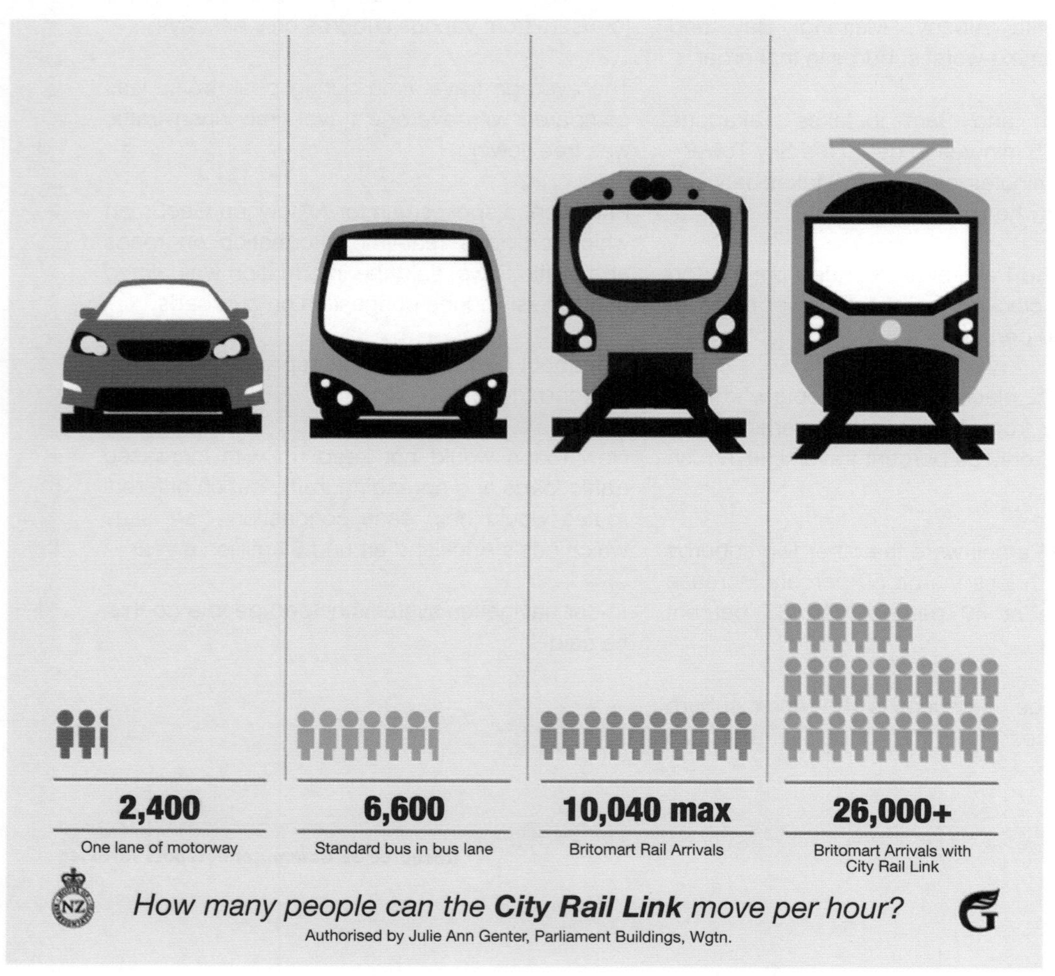

Resource 6G Transport mobility

ISBN: 9780170425285

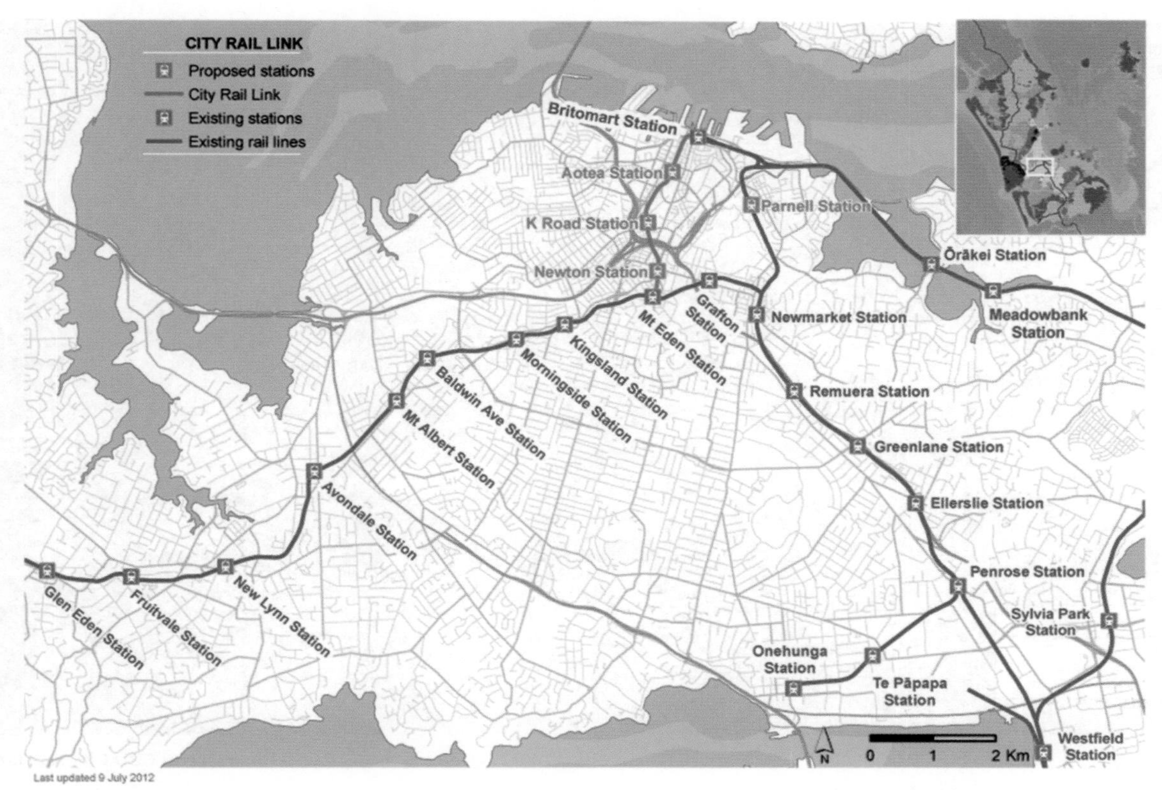

Resource 6H City Rail Link route

As shown in **Resource 6J**, the CRL will dramatically reduce travel times to and through the city centre, and people will have rail access to more parts of the city centre. For example, a public transport trip from New Lynn to the future Aotea Station will reduce by 55 percent from 51 to 23 minutes, while a public transport trip from Panmure to Newton will decrease by 33 percent from 40 minutes to 27 minutes.

The CRL will facilitate new commercial and residential development, and access to employment and educational opportunities not just for the city centre, but for all communities on the rail network. For example, Manukau and New Lynn town centres become more attractive as places to live and work because of the improved rail access to the city centre and across the network. The Auckland Council sees the CRL as a key enabler of increasing employment in the city centre and metropolitan centres on the rail network.

The CRL will help address congestion in the central city road network and enhance the ability of road corridors to handle the number of buses moving to/from the city centre, which, if left unaddressed, would limit the growth of the city centre and Auckland. This supports the planned refocusing of the bus network in outer areas to act as feeders to the rapid transit network. In addition, the CRL will improve transport choices for Aucklanders and reduce the environmental impact of the transport system.

Significant redevelopment has resulted from the initial upgrade of Auckland's rail network, including the double-tracking of the western line and the reopening of the Onehunga line, especially around Britomart, Newmarket and New Lynn stations, demonstrating a positive market response to investments in rail. The CRL will focus growth around Auckland's rail lines, supporting a more intensive city centre; the regeneration of traditional town centres; and the further development of newer centres, such as Sylvia Park.

The Auckland Council will underwrite the cost of protecting the City Rail Link route, acquire properties and prepare an updated business case for the CRL, compared with alternative options. The CRL, together with the purchase of new trains and improvements to the rail network, is estimated to cost $2.4 billion, and new funding tools to help pay for this project will be required. Auckland Council and Auckland Transport are developing a business case to support the funding and implementation of the CRL.

The CRL proposal is being developed as part of an integrated land-use and multi-modal transport approach, which includes: developing an integrated multi-modal package to optimise the accessibility of the city centre by all modes of transport. This includes improvements to the city centre bus network to address emerging capacity issues on key bus corridors, and to deliver customers closer to key city centre destinations identifying the potential for more housing and employment around railway stations and their catchments, and ensuring that land-use planning rules support this working with private sector partners to develop exemplar transit-oriented development projects around both the CRL and suburban railway stations reconfiguring bus services to act as feeders to rail at interchanges such as New Lynn, Onehunga, Manukau and Panmure providing additional park and ride sites to allow access to rail in locations without good public transport options.

ISBN: 9780170425285

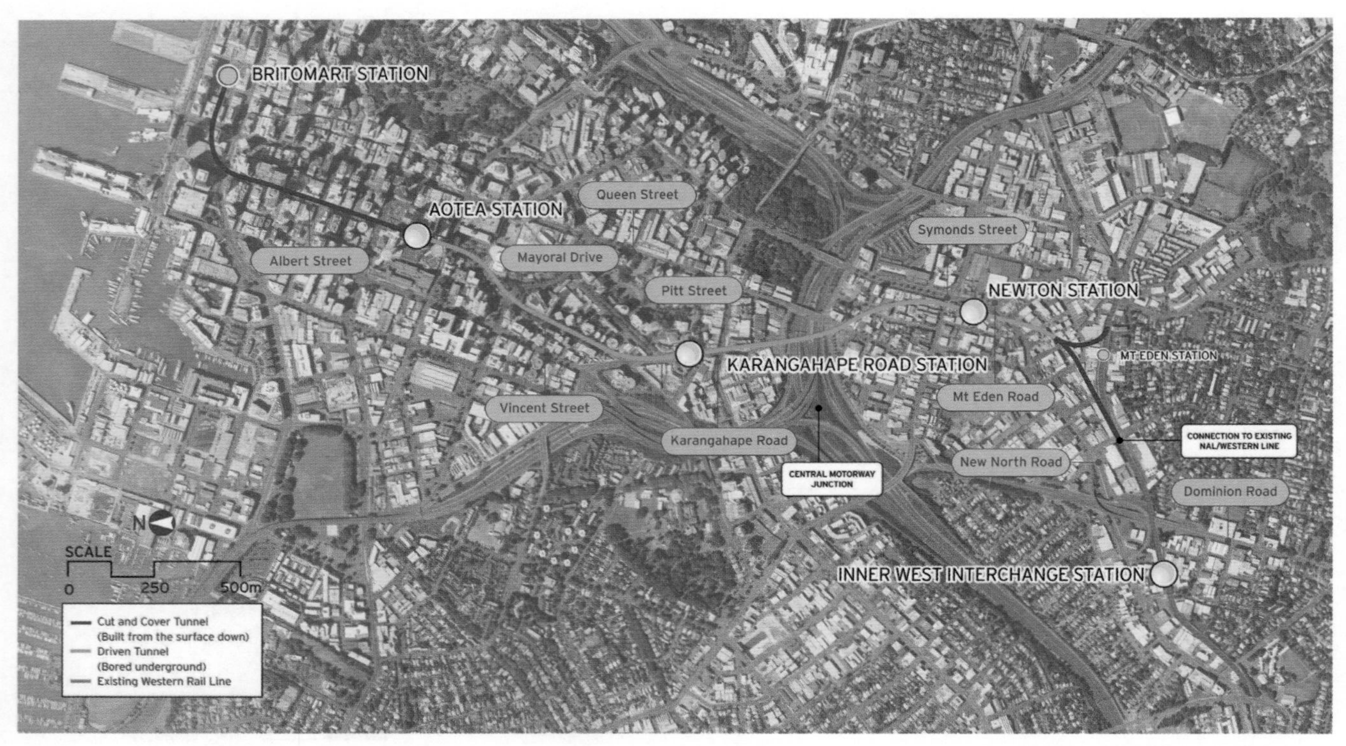

Resource 6I CBD City Rail Link route

Travel Time to CRL Station

From	To	Travel by train/bus (minutes) – Current	Travel by train/bus (minutes) – Future	Travel by train/bus (minutes) – Difference	% Improvement in Travel Time
Papakura	Aotea Station	66	54	12	18%
Manurewa	Aotea Station	57	47	10	18%
Papatoetoe	Karangahape Station	52	32	20	38%
Swanson	Karangahape Station	74	40	34	46%
Henderson	Aotea Station	59	35	24	41%
New Lynn	Britomart Station	35	27	8	23%
Kingsland	Karangahape Station	38	6	32	84%
Manukau	Karangahape Station	55	43	12	22%
Panmure	Karangahape Station	35	21	14	40%
Glen Innes	Aotea Station	25	15	10	40%
Onehunga	Aotea Station	40	31	9	23%
Ellerslie	Karangahape Station	36	17	19	53%
Newmarket	Aotea Station	20	11	9	45%
Karangahape Station	Britomart Station	18	6	12	67%

Resource 6J CRL travel times

ISBN: 9780170425285

Transmission Gully

Recognising the importance of moving freight and people between New Zealand's five largest cities to facilitate economic growth, the Government recently prioritised the construction of seven roads of national significance.

Of these, the Transmission Gully motorway project involves the construction of a 26-kilometre motorway link between Linden and MacKays Crossing, north of Wellington. The project, which will become an essential section of the Wellington Northern Corridor from Levin to Wellington, is considered a project of national significance that aims to 'reduce traffic congestion, improve safety and support economic growth in New Zealand'. The new motorway will also become part of State Highway 1 (SH1) that runs the length of New Zealand from Cape Reinga in the north to Bluff in the south.

Like all large-scale projects, the Transmission Gully motorway project is controversial and, as a result, has aroused widespread public concern and interest.

Q1 The Transmission Gully environment

Learning Activities

Applying a geographic concept: Environments

Environments may be natural and/or cultural. They have particular characteristics and features, which can be the result of natural and/or cultural processes. The particular characteristics of an environment may be similar to and/or different from another.

The Transmission Gully motorway project will provide an inland state highway alternative to the existing coastal route. The new motorway will mainly pass through rural land and will involve the cutting of 6.3 million cubic metres of material and the infilling of 5.8 million cubic metres. The new route will involve 112 stream crossings, requiring culverts and bridges to be constructed and the permanent realignment of several streams. Otherwise, land within the proposal area is 'highly modified, comprising mostly pasture, with some areas of regenerating native bush and exotic forestry'.

Refer to the definition of environments above and **Resources 7A – 7D** to answer this question.

a Complete the précis map on the following page by accurately locating and labelling the following features:

- **i** existing SH1 (coastal route)
- **ii** proposed Transmission Gully route (between Linden and MacKays Crossing)
- **iii** areas of native and exotic forestry east of the proposed inland route
- **iv** areas of urbanisation west of the proposed inland route.

b Calculate the distance of the existing SH1.

____

ISBN: 9780170425285

Learning Activities

Key

ISBN: 9780170425285

Learning Activities

c Describe the natural environment of the area surrounding Transmission Gully.

d Compare the topography of the existing coastal route to that of the Transmission Gully route.

e Identify three advantages the new inland route offers over the existing coastal route.

ISBN: 9780170425285

Q2 Mitigating the effects

Learning Activities

Applying a geographic concept: Perspectives

Perspectives are ways of seeing the world that help explain differences in decisions about, responses to, and interactions with environments. Perspectives are bodies of thought, theories or worldviews that shape people's values and have built up over time. They involve people's perceptions (how they view and interpret environments) and viewpoints (what they think) about geographic issues. Perceptions and viewpoints are influenced by people's values (deeply held beliefs about what is important or desirable).

The degree to which the developers of the Transmission Gully motorway project mitigate the effects on the environment will be subject to people's perceptions towards the project and the value they have placed on the Transmission Gully environment.

a Give reasons why this project is nationally significant, rather than just for the Wellington region.

ISBN: 9780170425285

Learning Activities

Refer to the definition of perspectives on the previous page and **Resources 7E** and **7F** to answer this question.

b Complete the table below identifying one positive and one negative social, economic and environmental effect from the Transmission Gully motorway project.

	Positive	**Negative**
Social		
Economic		
Environmental		

ISBN: 9780170425285

Learning Activities

Applying a geographic concept: Sustainability

Sustainability involves adopting ways of thinking and behaving that allow individuals, groups and societies to meet their needs and aspirations without preventing future generations from meeting theirs. Sustainable interaction with the environment may be achieved by preventing, limiting, minimising or correcting environmental damage to water, air and soil, as well as considering ecosystems and problems related to waste, noise, and visual pollution.

The Transmission Gully motorway project aims to sustain the potential of natural and physical resources for future generations while meeting the growing transportation needs of the Wellington region.

Refer to the definition of sustainability above and specific information from throughout the chapter to support your answer.

a Evaluate the extent to which the concept of sustainability has been applied to the Transmission Gully motorway project.

ISBN: 9780170425285

Learning Activities

ISBN: 9780170425285

Learning Activities

ISBN: 9780170425285

Resources

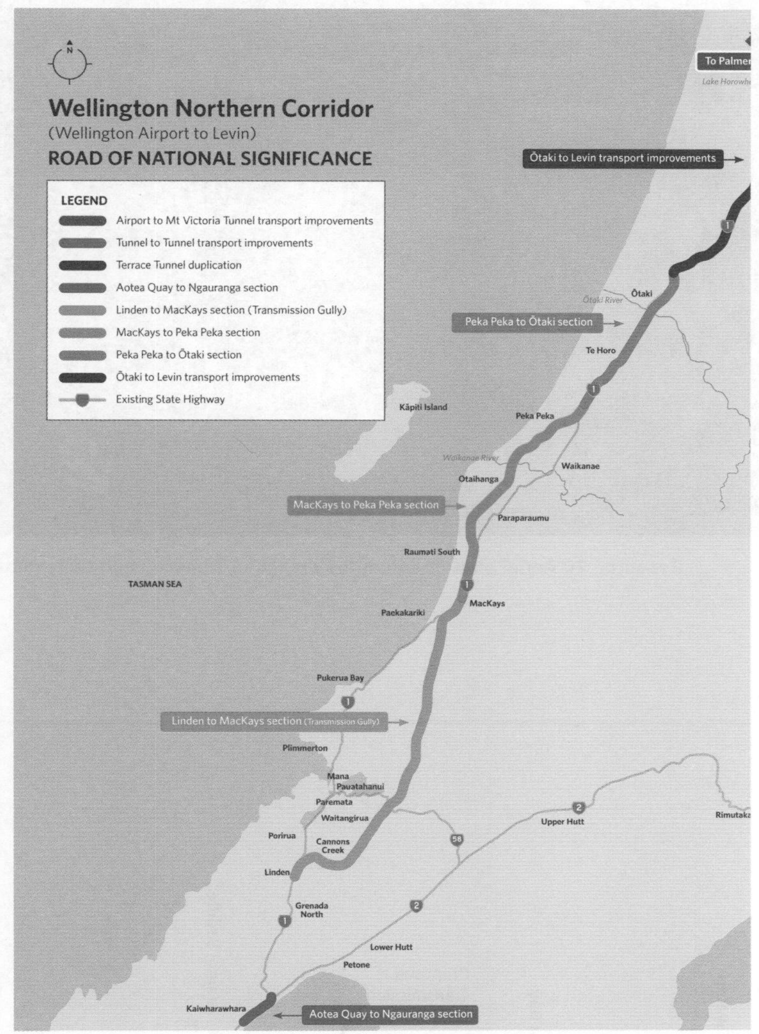

Resource 7A Wellington Northern Corridor location map

Resource 7B Horokiri Stream (left) and Wainui Saddle (right)

ISBN: 9780170425285

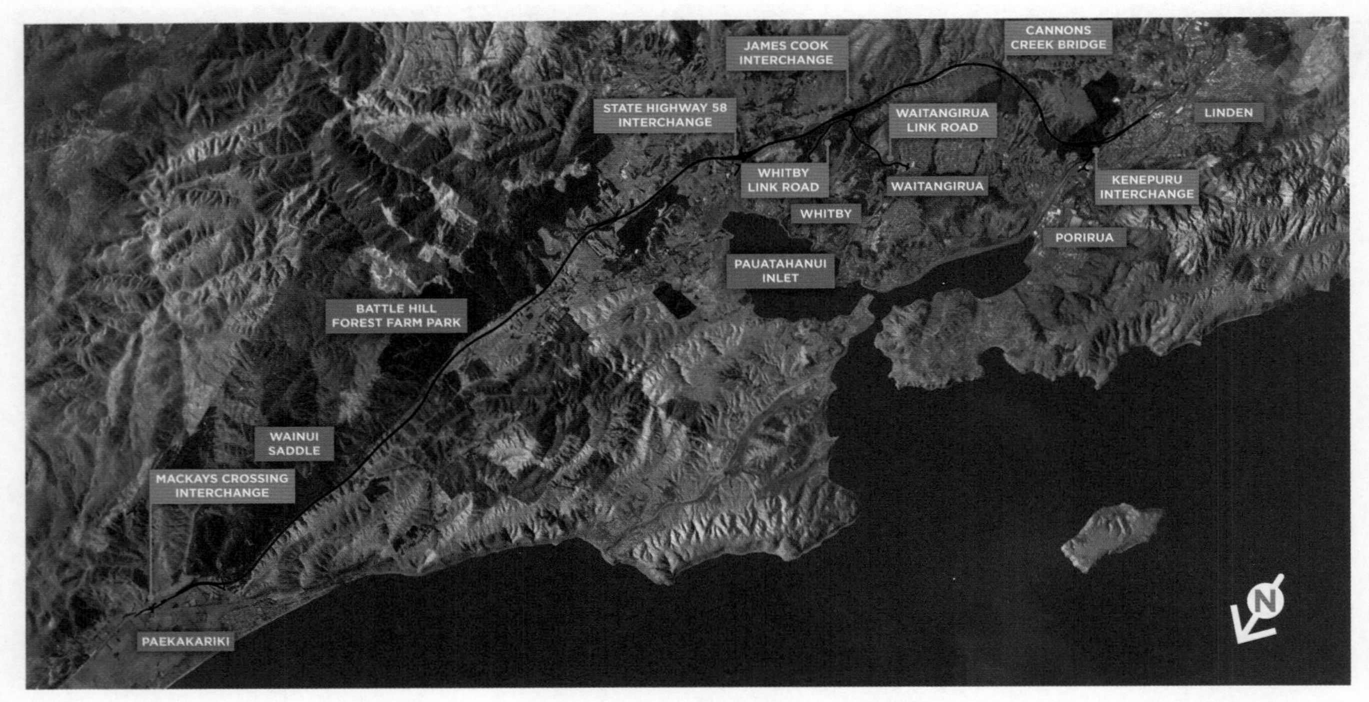

Resource 7C Aerial photograph of the proposed Transmission Gully route

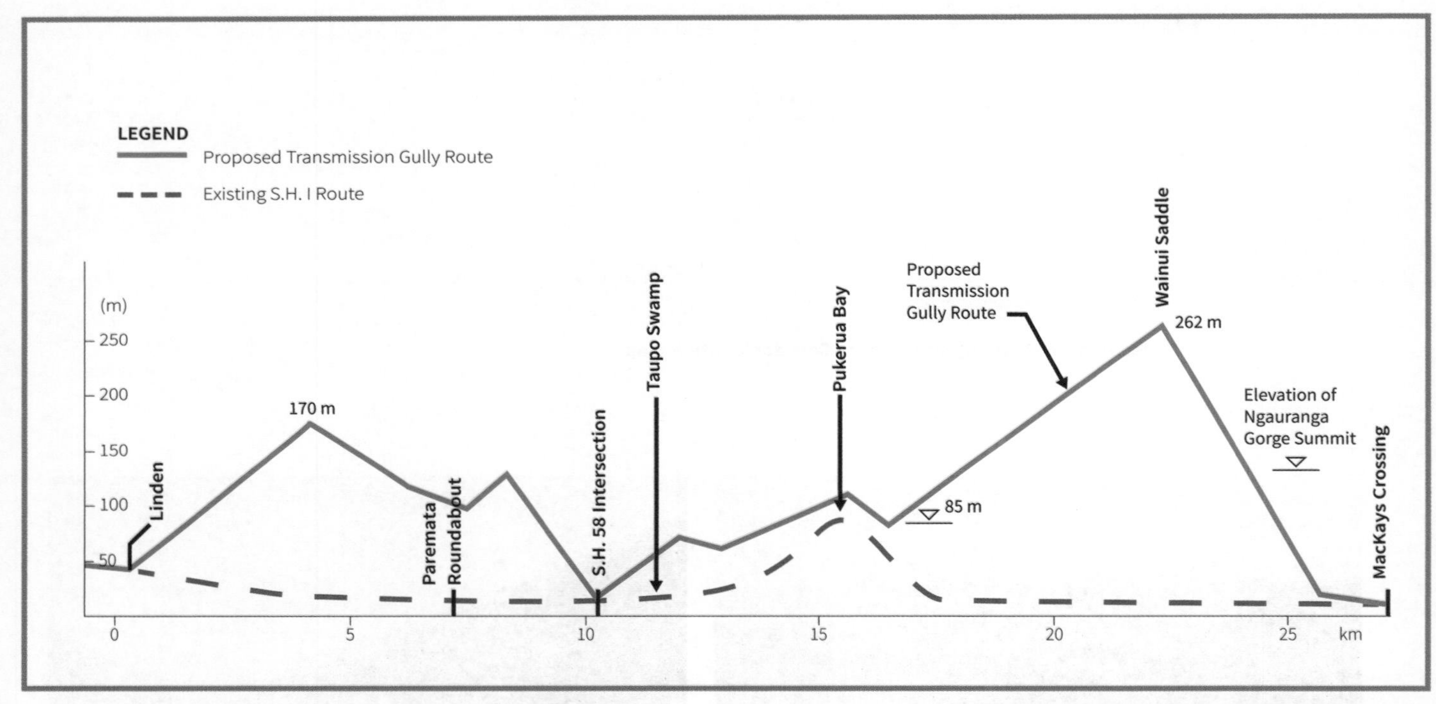

Resource 7D Transmission Gully route transect

ISBN: 9780170425285

Environmental Impact Assessment

Air quality

Construction of the project (particularly the earthworks and concrete batching plant) has the potential to generate nuisance dust. This will be managed through measures that are outlined in the Construction Air Quality Management Plan and include measures such as water sprays. The effect of operational vehicle emissions on air quality is also assessed. The assessment concludes that on completion of the project there will be an overall reduction in public exposure to vehicle emissions on a regional basis primarily due to the removal of vehicles from the existing coastal route, and reduced congestion.

Tangata whenua

The protection of stream habitats and resident native fish species is the key matter of interest to tangata whenua iwi (Ngati Toa) during both the construction and operational phases of the project. Parts of the project alignment and surrounds are customary food-gathering areas, and consequently there are areas of historical and cultural significance that have been recognised and provided for. Of particular interest to iwi are the direct and indirect effects of construction on water bodies, the most significant of which is the potential for increased levels of sediment entering waterways. Operational effects include discharge of stormwater from the road surface to streams and eventually to the Porirua Harbour, and effects on water and habitat quality.

Landscape and visual

The scale of the project means that it will create a significant change to the environment, and the scale of adverse visual and landscape effects varies as the road traverses through the landscape. The approach to designing the road and all associated works has been to avoid effects as far as practicable and to manage the remaining adverse visual and landscape effects. There is also the potential for positive visual effects for users of the road who will travel through the bold natural landscapes that are largely inaccessible at present.

Social effects

Construction and operation of the project has the potential to generate adverse social effects such as noise-related effects and general community disruption. The Construction Environment Management Plans will introduce measures to manage potential project effects and these will be finalised and implemented in consultation with local communities and stakeholders. This flexible approach is proposed so that any adverse effects on the social environment in nearby communities can be understood and managed in consultation with the community.

Positive social benefits of the project include reduced traffic, especially heavy vehicles, passing through the coastal communities, with associated noise reductions, improved air quality outcomes, and improved local ease-of-access and community cohesion as a result.

Freshwater ecology

As discussed above, sediment runoff from the earthworks has the potential to adversely affect freshwater habitats and species. The potential ecological impact of sediment runoff is assessed to be negligible, based on existing knowledge of stream environments and predicted sediment runoff levels. During construction, the earthworks areas are likely to increase sediment levels in streams — particularly during large rainfall events. The assessment concludes the ecological effects will be minor because the freshwater species existing in these streams currently are able to tolerate temporary increases in sediment levels. The steepness of the streams combined with the increased stream flows during rain events means that sediment is rapidly transported downstream rather than being deposited on stream beds.

Resource 7E Environmental Impact Assessment (EIA)

Transmission Gully impact: weighing up the evidence

Pockets of native bush and exotic trees will be hacked down to make way for Transmission Gully, forcing birds, bats and lizards out of their natural habitats.

Experts on one side say the 27km four-lane highway will actually benefit the environment in the long run. But opponents paint a grim picture of uncertainty.

Gully backers say it will dramatically reduce journey times, reduce crashes, boost the region's economy, reduce sediment entering Porirua Harbour in the long term and provide a secure route to the capital should disaster strike.

The project's final planning phase, after decades of debate, is taking place in a Wellington courtroom where lawyers in black suits grill experts to see if there are any holes in their evidence.

While many wrinkles have been ironed out at expert conferencing behind closed doors, some issues remain for the likes of Kapiti Coast District Council, Greater Wellington regional council and the Conservation Department.

Just how much sediment will pour into the already choked Porirua Harbour, including the nationally significant Pauatahanui Inlet, is unknown, and whether measures to enable threatened native fish to survive in affected streams is also unknown, objectors say.

But the New Zealand Transport Agency says it will achieve its goal of "no net loss" to the environment by either making up for loss where it occurs (mitigation) or making improvements to streams or forests away from the affected area (off-set mitigation). Stringent conditions, which are still evolving, will also ensure the environment is protected during construction and after the road is completed.

The main alignment of the alternative inland route was tweaked so it would mainly carve through farmland. However, streams and pockets of bush would still be casualties, ecologist Stephen Fuller said.

"Assuming the proposed mitigation is put in place, there will be a reduction in adverse effects over time, to the point where most effects are considered to be neutral."

He appeared before the board of inquiry this week on behalf of applicants NZTA, Porirua City Council and Transpower.

The board can either kill the project or approve it with conditions. If approved in June, construction could start in 2015 and be completed by 2021.

The safer and faster road into Wellington would come at a cost of 40 hectares of native vegetation being "permanently lost" during construction, Dr Fuller said. A further 80ha of native vegetation within the project boundary may be lost or modified by earthworks. But the retirement and revegetation of 627ha of land would make up for the ecological losses, he assured the board this week.

Marine ecologist Sharon De Luca told the board the "life-supporting capacity" of Porirua Harbour would not be affected.

Additional sediment accumulation would have "some small" adverse effects on the ecological values and functioning of the harbour, but the 50mm of sediment expected to accumulate due to the project would settle in areas with low ecological values, she said on behalf of NZTA.

However, Conservation Department experts are dubious about the calculations and concerned about the effects on the inlet.

The estuary was specifically recognised by Environment Minister Nick Smith in his decision to refer the roading proposal to an independent board of inquiry, under new rules to fast-track projects of national significance.

Ecologist Helen Kettles said the effects could be more severe than what NZTA says and mitigation would not adequately protect its significant ecological values.

"Ultimately if reductions in sediment inputs into the Pauatahanui Inlet are not actioned it may well become a brackish swamp in as little as 145 years."

Department lawyer Shona Bradley told the board it was not opposed to the $1 billion project, but wanted conditions addressing the short and long-term effects.

Kapiti Coast District Council focused on effects in its patch and engaged Mike Joy, a Massey University senior lecturer in ecology and environmental science, to give his opinion on the extent of damage to Te Puka Stream.

It contains significant populations of threatened native migratory fish – longfin eel, koaro and redfin bully – and will be significantly modified and diverted.

He had "little faith" that experimental measures to enable fish to migrate would work and called for intense monitoring to ensure fish did not die en masse.

More than 10 kilometres of streams would be lost, modified or diverted by the project. There would be 15 bridges and 125 culverts. Of those 140 water crossings, 35 would require a fish passage to enable migration. A further 10 waterways could require a fish passage.

Ecologist Vaughan Keesing told the board on behalf of NZTA that mitigation measures would ensure all 17 fish species in the seven affected catchments survived.

"The adverse effects of the project on most streams will be ecologically significant; however, I consider that the proposed mitigation is sufficient to ensure that the functional integrity of the waterways is maintained, and that no fish species are lost."

The board also heard Transmission Gully could make the Ngauranga Gorge bottleneck worse if the proposed 6.4km Petone-Grenada link road is axed.

Greater Wellington senior transport planner Natasha Hayes confirmed no modelling had been done on the congestion impact on Wellington traffic if the link road, forecast to carry 25,000 vehicles a day by 2026, wasn't built.

Transmission Gully Action Group chairwoman Grace Osvald warned Transmission Gully was the only option left to ease congestion after Greater Wellington adopted a hearing panel's decision in 2006 to abandon plans to upgrade the existing coastal route.

"If resource consent for this project is declined, what then? We know the coastal upgrade option is unconsentable and unacceptable, the only option then potentially becomes 'do nothing'."

The hearing will take a break next week for more conferencing and hopefully more resolutions on conditions. It resumes on March 5.

Resource 7F Transmission Gully impact: weighing up the evidence

ISBN: 9780170425285

Rwanda: Land of a Thousand Hills

A little less than three decades on from the Rwandan genocide, the Republic of Rwanda is re-emerging as Africa's leader in sustainable tourism development. With breathtaking landscapes, diverse flora and fauna, and a rich traditional culture, Rwandans have embraced sustainable tourism and conservation practices as a way of improving the well-being of its rural communities. As one of only three countries in the world where the critically endangered mountain gorillas live, Rwanda has been able to successfully take advantage of its abundant wildlife, culture and the network of national parks to attract more ecologically aware visitors to the country than ever before.

Q1 Rwanda's national parks

Learning Activities

Applying geographic concepts: Location and accessibility

Location: The position of something that can be given in absolute terms, or in relation to other objects. Location can be an advantage or a constraint.

Accessibility: A measure of the ease of movement of people or ideas. The greater the accessibility, the greater the potential for change.

Refer to the definitions of location and accessibility above and **Resources 8A – 8C** to answer the following questions.

a Describe the location and accessibility of Rwanda's network of national parks. Use geographic skills such as distance, direction, latitude and longitude, and relief interpretation, as well as specific information from the resources, to support your answer.

ISBN: 9780170425285

Learning Activities

b Use an appropriate statistical mapping technique to display the number of park visits to each of Rwanda's three main national parks.

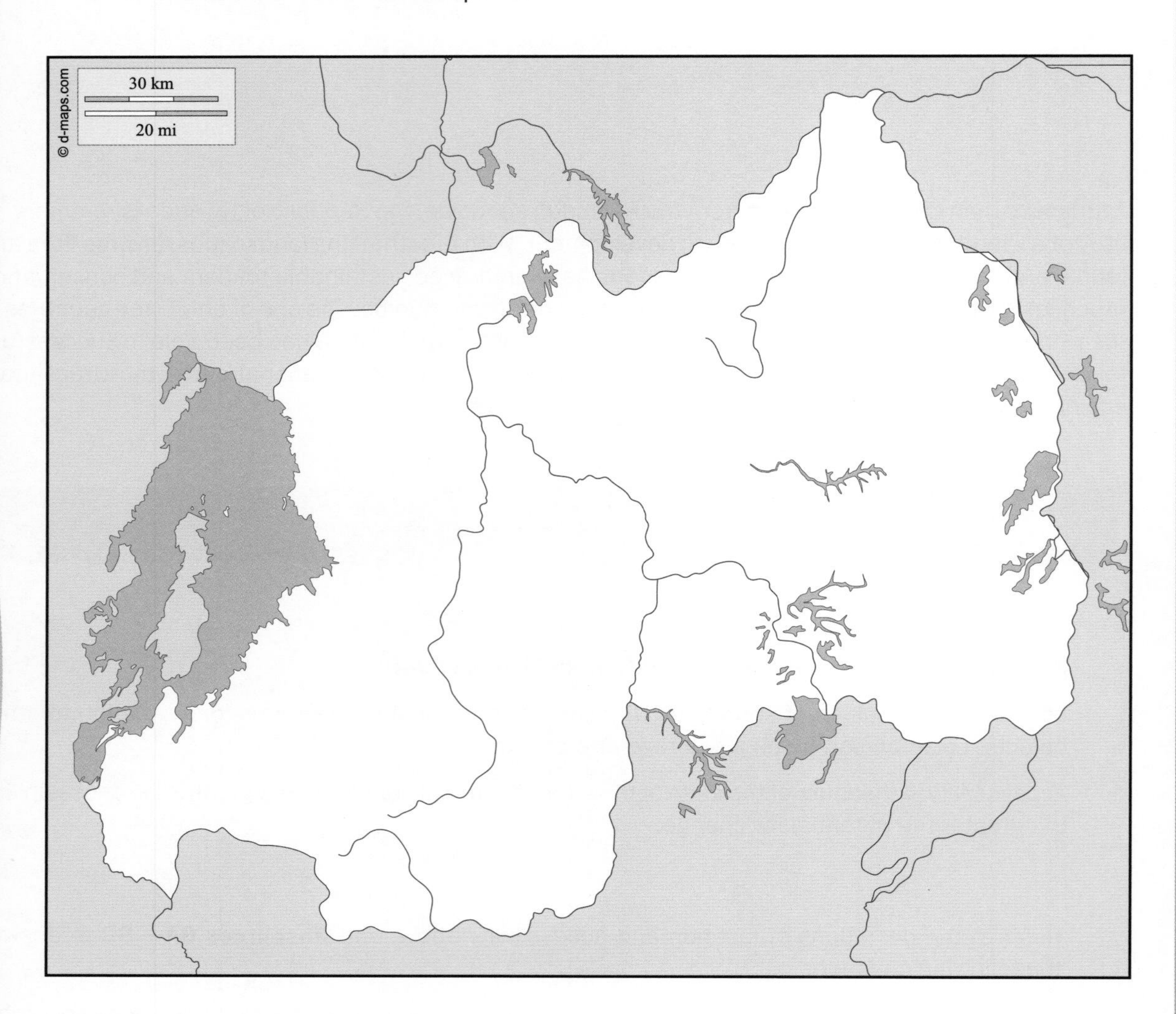

c With reference to the statistical map you have drawn, explain in detail the pattern of visits to Rwanda's national parks. Tip: use PQE.

ISBN: 9780170425285

Q2 Tourism growth in Rwanda

Learning Activities

Applying a geographic concept: Change

Change involves any alteration to the natural or cultural environment. Change can be spatial and/or temporal. Change is a normal process in both natural and cultural environments. It occurs at varying rates, at different times and in different places. Some changes are predictable, recurrent or cyclic, while others are unpredictable or erratic. Change can bring about further change.

Refer to the definition of change above and **Resources 8D** and **E** to answer the following questions.

a List in order the origin where most of Rwanda's non-business-related visitors originated in 2014.

i Rank 1

ii Rank 2

iii Rank 3

b Draw an appropriate graph below, using the information from the table in **Resource D**, to show the purpose of travel by region of origin in 2014. Include all appropriate graphing conventions. Tip: use SALTS.

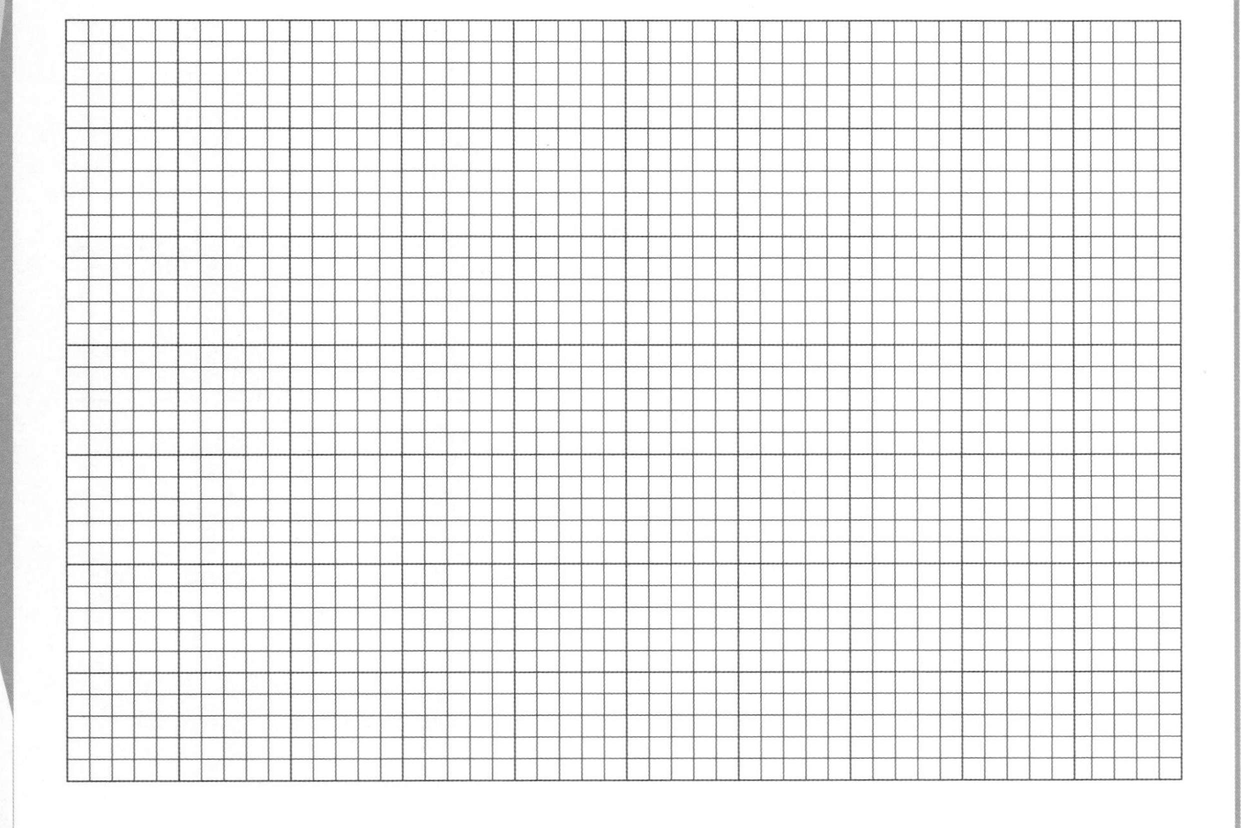

ISBN: 9780170425285

Learning Activities

c Contrast the main purpose of Africans visiting Rwanda to that of other regions. Tip: use PQE.

d Describe the change in the contribution of travel and tourism to the GDP of Rwanda 2008–2017.

ISBN: 9780170425285

Q3 The consequences of tourism

Applying a geographic concept: Interaction

Interaction involves elements of an environment affecting each other and being linked together. Interaction incorporates movement, flows, connections, links and interrelationships. Landscapes are the visible outcome of interactions. Interaction can bring about environmental change.

Refer to the definition of interaction above and **Resources 8F – 8K** to answer the following questions.

a Fully explain an important social, economic and environmental consequence of tourism growth in Rwanda. Use specific information from the resources to support your answer.

One social consequence:	
One economic consequence:	
One environmental consequence:	

ISBN: 9780170425285

Learning Activities

b Choose and circle one of the following geographic concepts relevant to the growth of travel and tourism in Rwanda.

Change Sustainability Interaction

Comprehensively analyse how this geographic concept is relevant to the growth of tourism in Rwanda. Use specific information from the resources to support your answer.

ISBN: 9780170425285

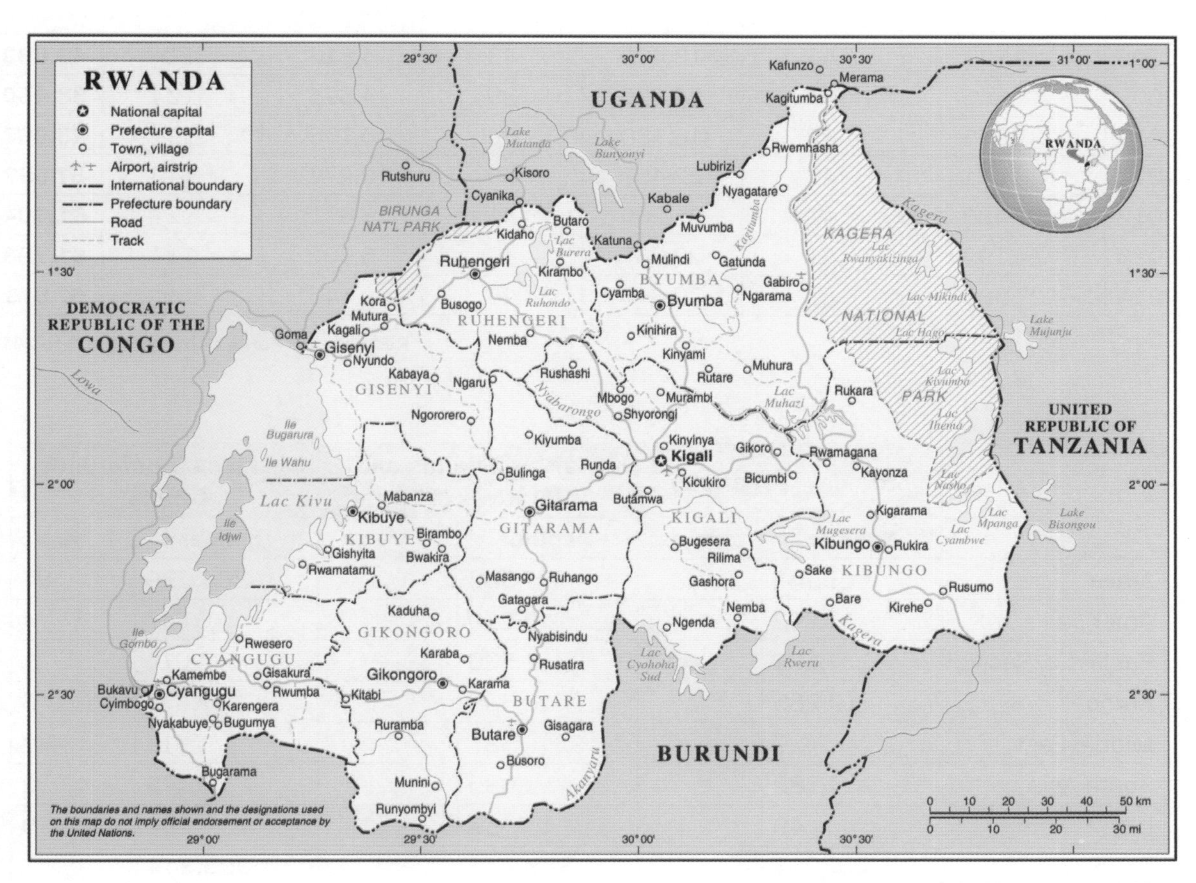

Resource 8A United Nations map of Rwanda

Resource 8B Satellite image of Nyungwe Forest National Park and Lake Kivu (top left)

ISBN: 9780170425285

Year	Birunga - Volcanoes National Park		Akagera (Kagera) National Park		Nyungwe National Park		Total
	Number	%	Number	%	Number	%	
2008	19,783	46	18,490	43	4,810	11	**43,083**
2009	18,865	49	14,890	39	4,695	12	**38,450**
2010	23,372	52	16,180	36	5,755	13	**45,307**
2011	26,821	47	22,457	39	8,274	14	**57,552**
2012	28,483	47	25,200	41	7,621	12	**61,304**
2013	25,199	41	29,687	48	6,902	11	**61,788**
2014	27,885	41	30,846	45	9,140	14	**67,871**

Resource 8C Parks visits from 2008 to 2014

	Holiday/Vacation		Visiting Friends and Relatives (VFR)		Business/Conference/ Official	
	Number	%	Number	%	Number	%
Africa	65,747		377,583		364,690	87
America	13,792		6,210		6,817	4
East Asia/Pacific	4,131		1,500		4,076	1
Europe	19,285		10,538		12,377	6
Middle East	427		587		711	0
South Asia	1,782		1,635		6,479	2
UN	49		58		194	0
Total	**105,213**		**398,111**		**395,344**	

Resource 8D Purpose of visit by region, 2014

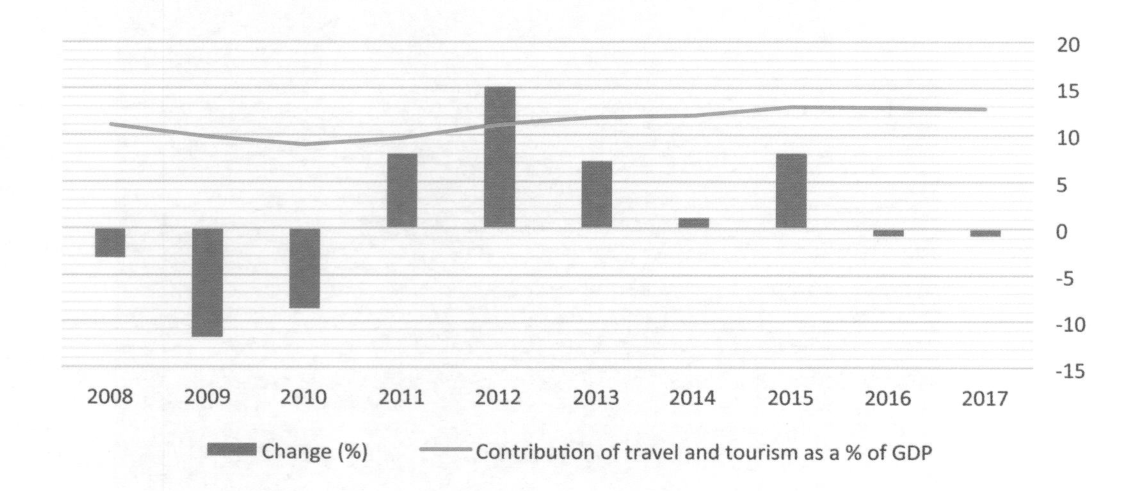

Resource 8E

ISBN: 9780170425285

Resource 8F A trekking path on the slopes of Volcanoes National Park leading into the thick forest towards the gorilla families

Resource 8G A group of tourists climbing up the bamboo forest slopes of Virunga Mountain in Volcanoes National Park

Resource 8H A tourist and family of mountain gorillas

Resource 8I A family of mountain gorillas

Annual visitor numbers continue to increase at Akagera

Akagera National Park in eastern Rwanda has seen a steady increase in visitor numbers since Akagera Management Company (AMC), a joint partnership between the Rwanda Development Board and African Parks, assumed management of the park in 2010.

In 2011, AMC reported a total of 20,657 visitors to Akagera, which was a substantial 35% increase in the park's visitor numbers from 2010. In 2012, the park saw 23,048 visitors, an increase of 12% from 2011 and an overall 24% increase over the past two years.

In addition, 50% of the visitors were Rwandan nationals, demonstrating the extent to which Akagera has become a source of national pride within Rwanda. The entrance fees and activities alone produced over $500,000 in revenue. This growth in tourism receipts coupled with the steady increase in visitor numbers means that Akagera is well on track to becoming economically self-sustaining within the next 5-7 years.

Akagera National Park has been widely recognized as having the potential to become one of Rwanda's premier tourism attractions. Since AMC took over management in 2010, the park has seen numerous improvements in infrastructure, including the development of new road networks, the maintenance of existing roads, the development of new signage and maps and the introduction of boats for lake-based tourism activities.

In 2012, Akagera Management Company started and completed the Ruzizi Tented Lodge, which sits on the south-western side of Lake Ihema, the largest lake in the park. The lodge was started in January 2012 and completed for opening to the public in October 2012. The new tented lodge is the first of its kind in Rwanda and it is already attracting a good occupancy and generating income for the park.

During 2013, emphasis will be placed on developing the tourism experience, including the completion thereof a day visitor's centre at the main entrance to the park, which will be officially opened along with Ruzizi Tented Lodge in March 2013.

Resource 8J

ISBN: 9780170425285

How gorillas and local communities depend on each other

May 19, 2018

By Joseph Ondiek

There has an exciting resurgence in the population of Rwanda's famous mountain gorillas. Although their numbers once dwindled due to rampant poaching, conflicts, and habitat loss, the Volcanoes National Park in Rwanda has over the past years witnessed steady growth of the mountain gorillas.

This has widely been attributed to the country's conservation efforts. Rwanda's tourism industry is buoyed by the existence of these primates. The relationship between Rwanda and the gorillas is some unique form of interdependence.

Gorillas are Rwanda's most popular attraction

There might be a lot to see and feel in Rwanda. This is a country of picturesque rolling hills and dazzling beauty, captured by its rich culture and enticing dance, the Intore.

However, the mountain gorillas still remain Rwanda's flagship tourist attraction, as they bring most of the tourists to the country. Lodges like luxurious Bisate Lodge and campsites like Red Rocks Cultural Center in Musanze are built around the very concept related to gorilla tours.

New tourism-related organizations like the Virunga Community Programs, restaurants, tour companies and boutiques have sprung up to offer mountain gorilla-related services. It's in this regard that the mountain gorillas have undoubtedly contributed significantly to Rwanda's international recognition.

Saving deforestation and gorilla habitat loss

There are about 1,000 mountain gorillas left in the world. Half of these reside in the Virunga mountain range. The range is shared by Uganda, Rwanda and Democratic Republic of Congo (DRC).

Among the biggest threats to the existence of the gorilla is habitat loss. But Rwanda has made significant, and laudable, efforts in ensuring the size of the park is maintained. This has contributed to ensuring the gorillas are suitably protected. Perhaps, as a result, over 60 percent of gorillas in the Virunga massif reside in Rwanda.

Peace and prosperity has led to more mountain gorillas

Sadly, just some few years after Dian Fossey's relentless gorilla conservation efforts, conflicts in gorilla habitats deterred the growth in the population of the gorillas. But after the harrowing 1994 genocide, the country embarked on an ambitious plan to rebuild itself literally from the rubble.

Together with the rehabilitation of towns, cities, and communities, also the gorilla population began making a strong comeback. Since the 1994 genocide, the country has experienced unprecedented peace, political and social stability, and fewer human threats to mountain gorillas.

Restricted tourist activities to keep the gorillas both happy and healthy

Did you know that gorillas share 98 percent of their DNA with the human being? Consequently, gorillas notoriously have frail immune systems concerning interacting with humans. But Rwanda has done everything possible to prevent the spreading of airborne viruses, bacteria, and disease.

Tourists are not permitted to trek to the gorillas when they have cold or flu, and when they reach the gorillas can't stand closer than 10 meters (30 feet).

Basically, to keep the gorillas happy, tourists are only permitted to observe them for one hour before they return down the mountain. And the gorilla doctors also are always ready in case any of the gorillas happen to fall ill.

The gorilla permit costs fund directly sustainable community development, and this has further led to the primates' conservation.

In 2017, Rwanda doubled its gorilla trekking permits from $750 to $1500. Though the decision was met with some outcry, it was nobly done to raise more fund for conservation efforts and support the surrounding communities.

Previously, communities were provided $37 from each gorilla-trekking tourist for hospitals, schools, business development and clean water sources. Rwanda now provides surrounding communities $150 per tourist.

Before the institutionalized funds, the communities were among the biggest threats to the gorillas, since impoverished families would poach these gorillas, take food and trees from their habitat, or even accidentally kill them while poaching for other animals.

But now, with tourism dollars impacting directly these communities, the motivation to preserve them is greater, and also now reformed poachers can be found in a range of roles like porters, gorilla trackers, conservationists, and park rangers.

Resource 8K

ISBN: 9780170425285

Rwanda's tourism earnings seen up 25 pct in 2016 from last year – official

KIGALI, Sept 3 (Reuters) – Rwanda expects to boost its revenue earnings from tourism this year by 25.8 percent from 2015, helped by extra attractions including a new game park, an official told Reuters late on Friday.

The central African country famed for rolling green hills and treks to see endangered gorillas on the slopes of the Virunga Mountains, sees tourist earnings reaching $400 million this year, up from $318 million in 2015.

Francis Gatare, chief executive officer of state-run Rwanda Development Board (RDB), told Reuters that Rwanda wanted to maximise earnings from visitors "by giving several opportunities so that they [visitors] can increase the length of stay in the country."

A new national park called Gishwati-Mukura, the country's fourth, a "cultural village" in the capital Kigali, a vast new hospitality facility called the Kigali Convention Centre, and new adventure activities on Lake Kivu would drive revenue growth, Gatare said.

Rwanda is credited with rapid growth since the 1994 genocide that claimed the lives of 800,000 people but critics of incumbent President Paul Kagame say his authoritarian style undermines the potential for long-term political stability.

The nation's misty valleys and mountain gorillas are a strong magnet for tourists and visitor numbers in 2016 are expected to increase by 4 per cent up from 1.3 million in 2015.

Resource 8L

ISBN: 9780170425285

Sea Ice Express

Sea ice is frozen seawater that floats on the ocean surface. Covering millions of square kilometres, sea ice forms and melts with the polar seasons, affecting both human activity and biological habitats. In the Arctic, some sea ice remains frozen year after year, whereas Antarctic sea ice is seasonal, meaning it melts away and reforms annually. While both Arctic and Antarctic ice are of vital importance to the marine mammals and birds for which they are habitats, sea ice in the Arctic seems to play a more crucial role in regulating climate.

Since 1979, the extent of sea ice covering the Arctic in winter has decreased at a rate of 3 to 4 percent per decade. The cause of this decline is the subject of intense debate, however, an increase in global mean temperatures and climate variability is the likely cause of sea ice decline.

Q1 Natural environment of the Arctic region

Learning Activities

Applying a geographic concept: Environments

Environments may be natural and/or cultural. They have particular characteristics and features, which can be the result of natural and/or cultural processes. The particular characteristics of an environment may be similar to and/or different from another.

The vast Arctic region extends across the continents of North America, northern Europe and northern Asia, taking in eight countries and large expanses of ocean and sea in between. The terrestrial, freshwater and marine environments throughout this area exhibit considerable variation in climate, meteorology and physical geography.

a Refer to the definition of environments above and **Resources 9A – 9G** to analyse three characteristics of the natural environment of the Arctic region from the diagram below:

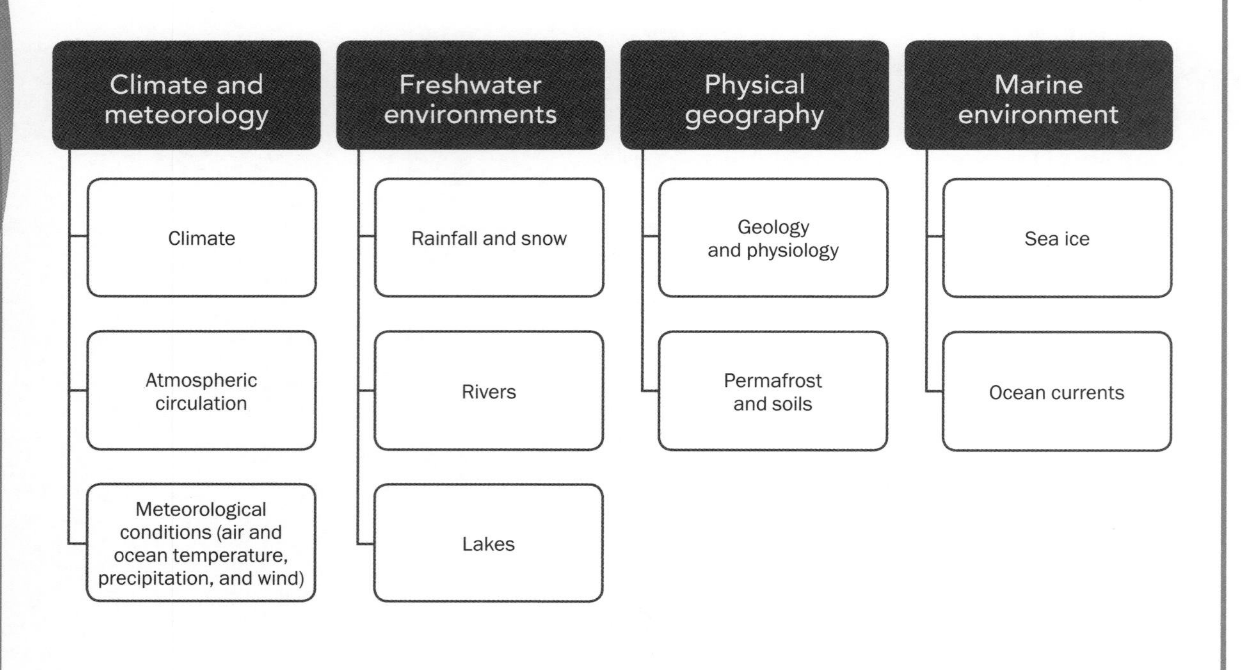

ISBN: 9780170425285

Learning Activities

i

ii

iii

ISBN: 9780170425285

Q2 Spatial and temporal patterns

Learning Activities

Applying a geographic concept: Patterns

Patterns may be spatial (the arrangement of features on the earth's surface) or temporal (how characteristics differ over time in recognisable ways).

Refer to the definition of patterns above and **Resources 9A**, **9H** and **9I** to answer this question.

a Describe, in detail, one spatial and one temporal pattern associated with the global distribution of sea ice. Tip: use PQE.

i Spatial pattern:

ii Temporal pattern:

b With reference to the ocean and eight nations located within Arctic Circle, describe the geopolitical composition of the Arctic region.

ISBN: 9780170425285

Q3 Interaction brings about environmental change

Applying a geographic concept: Interaction

Interaction involves elements of an environment affecting each other and being linked together. Interaction incorporates movement, flows, connections, links and interrelationships. Landscapes are the visible outcome of interactions. Interaction can bring about environmental change.

The link between Arctic sea ice extent, ocean currents, climate, and natural ecosystems in bringing about environmental change is well documented. However, what is less understood is the way the elements of the environment interrelate to bring about change.

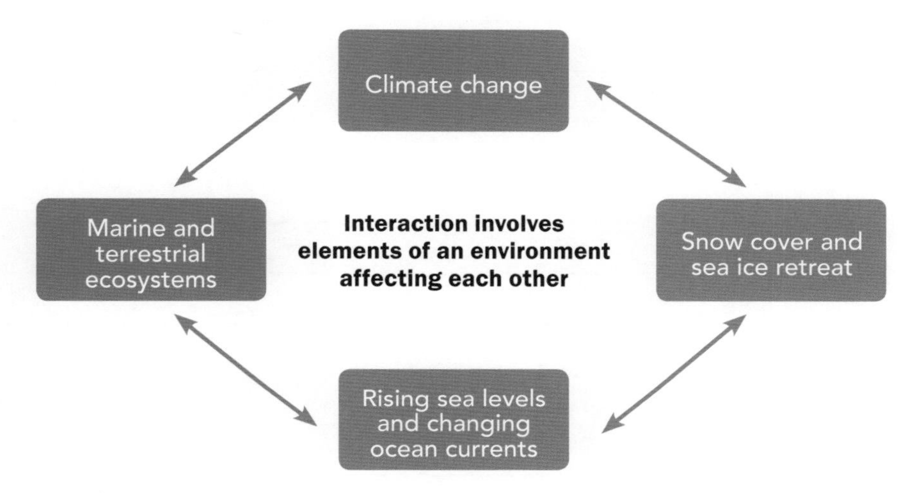

a Give reasons as to why the extent of sea ice in the Arctic region is of global significance.

ISBN: 9780170425285

Learning Activities

b Refer to the definition of interaction and **Resource 9J** to explain in detail how elements of the environment interact to bring about environmental change.

ISBN: 9780170425285

Learning Activities

ISBN: 9780170425285

Resources

The Arctic region, which covers an area of approximately 20,000,000 km^2, is located at the northernmost part of the Earth. Its topography is diverse consisting of ice-covered ocean, extensive glaciers and vast areas of snow and permafrost (frozen soil). The red line in the map indicates the 10°C isotherm in July (summer), which is commonly used to delimit the Arctic region, and the white area shows the average minimum extent of sea ice in summer as of 1975.

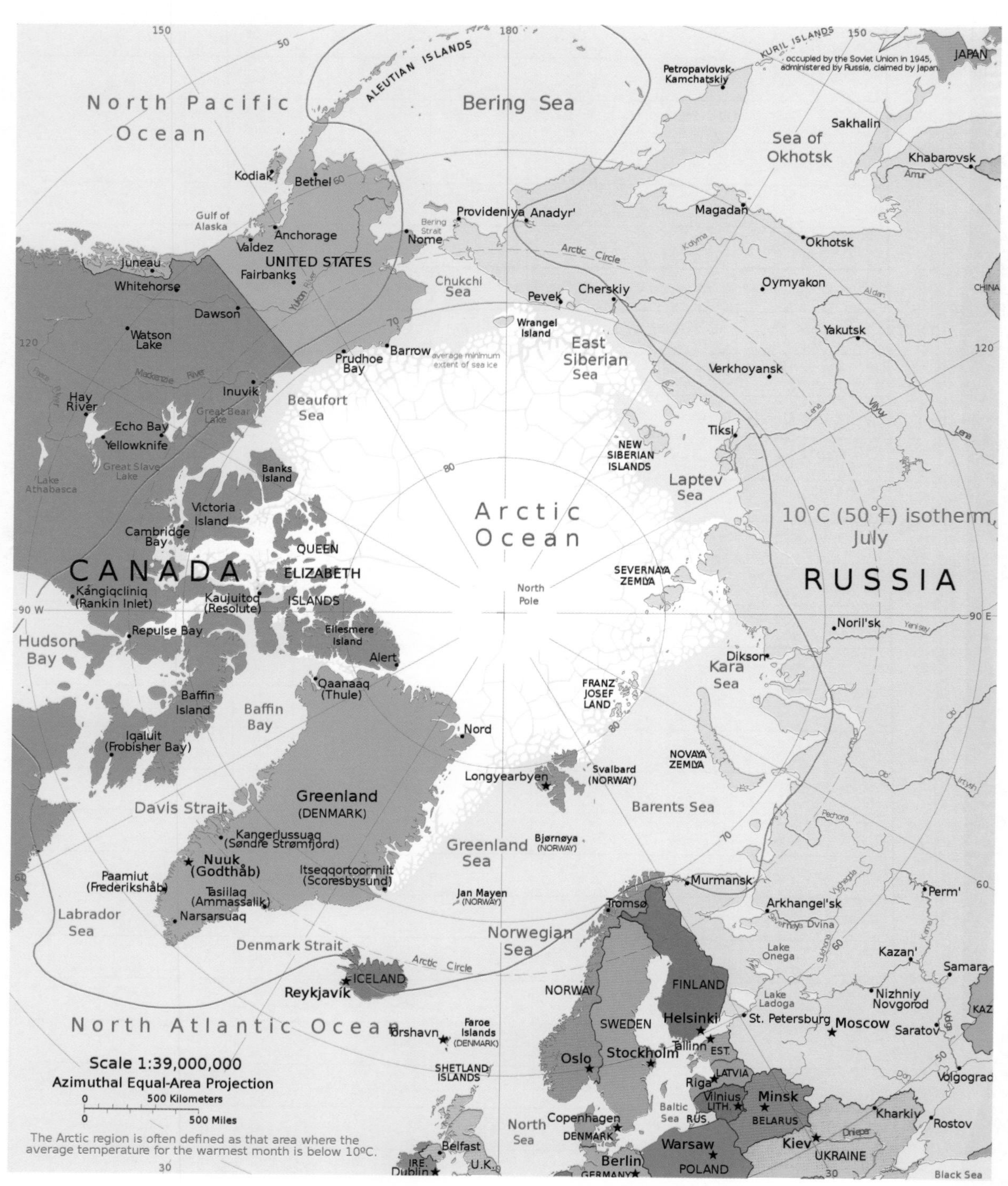

Resource 9A Relative location of the Arctic region

ISBN: 9780170425285

Sea ice is simply frozen seawater. In contrast to icebergs, glaciers, ice sheets and ice shelves, which all originate on land, sea ice forms, grows and melts in the ocean. Sea ice occurs in both the Arctic and Antarctic. In the northern hemisphere, sea ice can exist as far south as Bohai Bay, China (approximately 38°N), which is about 700 kilometres closer to the equator than it is to the North Pole. Sea ice grows during the winter months and melts during the summer months, but in cooler areas of the Arctic Ocean some sea ice remains all year.

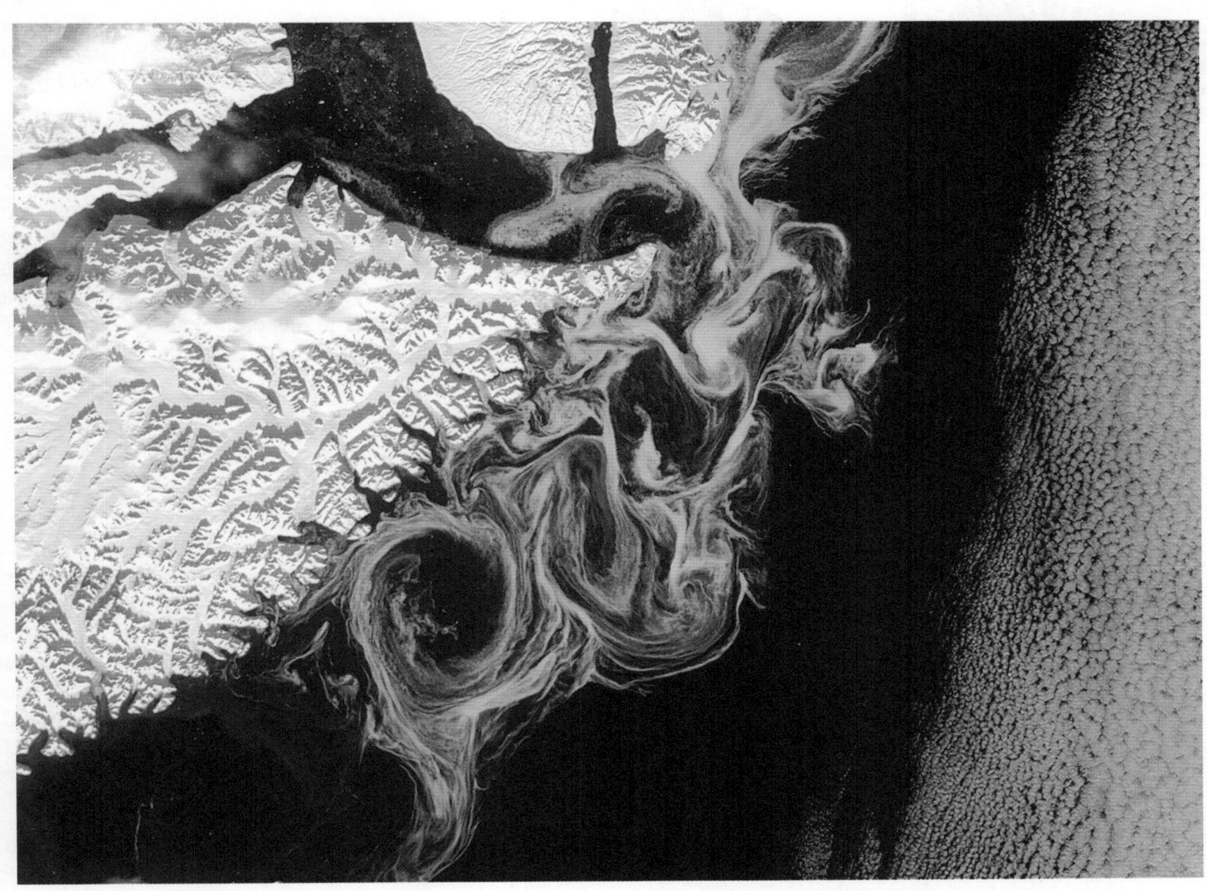

Resource 9B Swirls of sea ice off the coast of Greenland

ISBN: 9780170425285

Air Temperature Anomaly (degrees C) May 2010

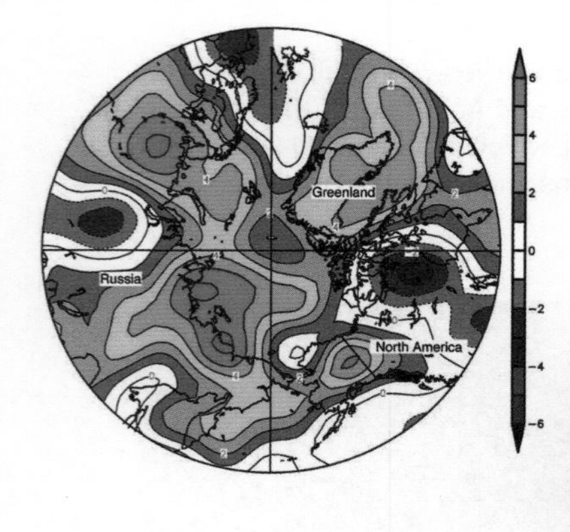

Sea Level Pressure Composite Mean (MB) June 2010

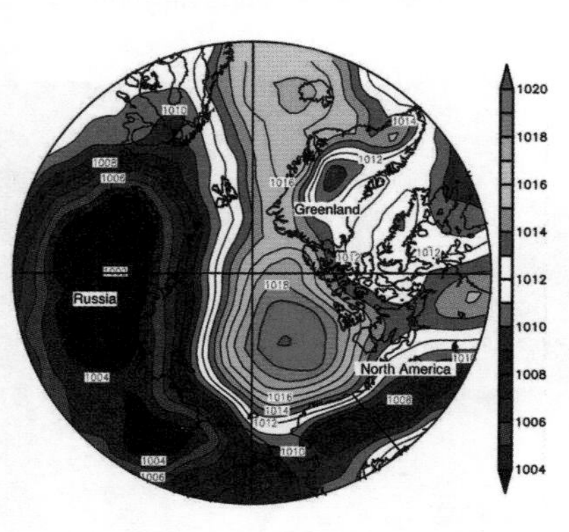

Resource 9C Arctic air pressure map

The Arctic region is characterised by low air temperatures. This is because, on an annual basis, the region receives less solar energy from the sun than other parts of the Earth. However, the amount of the sun's energy the Arctic region receives does vary greatly depending on the season. In the summer months (June-August), the Arctic receives higher levels of solar energy than any other place on Earth; while in the winter months (November-February) the region receives none of the sun's solar energy.

Annually, the Arctic region receives less solar energy than is lost to space by long-wave radiation. This is partly because a large portion of the solar energy that reaches the region is reflected back into space by extensive cloud, snow and ice cover. The ability for cloud, and snow and ice cover to reflect the sun's solar energy is commonly referred to as the 'albedo effect'.

Climate graph Nord, Greenland 81.59°N, 16.60°W

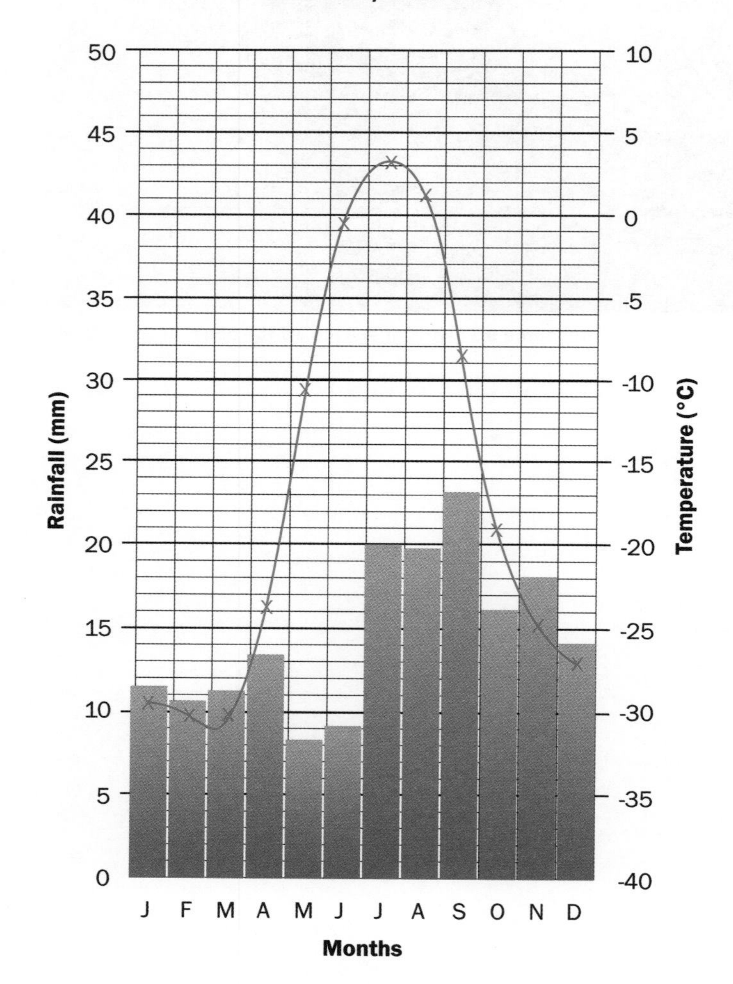

Resource 9E Climate graph

The seasonal climate of the Arctic is characterised by long cold winters and short cool summers. Some parts of the Arctic are covered by ice year-round, and nearly all parts of the Arctic experience long periods covered by ice (sea ice, glacial ice, or snow). Average January temperatures range from about −40 to 0°C but can drop below −50°C. Average July temperatures range from about −10 to 10°C while summer temperatures rarely rise above 15°C.

Climate conditions in the Arctic are divided into oceanic and continental subtypes. The Arctic oceanic climate is characterised by stormy winters but rarely fall below −7°C. Summers are cloudy and temperatures average a mild 10°C. Continental climates which characterise the inland areas of northern Canada, Greenland, Scandinavia, northern Russia and Alaska experience lower precipitation but greater extremes between summer and winter.

Most precipitation in the Arctic falls as snow, which accumulates as snowpack over the winter. Snowpack duration ranges from about 180 days to more than 260 days. High levels of solar energy reaching polar latitudes in spring results in rapid snowmelt. Spring runoff comprises 80-90% of the yearly total, and lasts only two to three weeks. Infiltration of this flush of water into the ground is constrained by the permafrost. Thus, spring melt water may flow over land and enter rivers, or accumulate into the many ponds and lakes characteristic of low-lying areas. Summer sources of water include late or perennial snow patches, glaciers, rain, melting of permafrost, and groundwater discharge.

ISBN: 9780170425285

The landscape of the Arctic region is extremely diverse. Glaciers (large masses of ice that flow under their own weight) form throughout the Arctic region where winter snowfall exceeds summer melting. Iceland is mountainous and volcanically active; half of its land area is devoid of vegetation and 11% of its land area is covered by glaciers. Northern Russia is characterised by lakes. The Russian Arctic west of the Ural Mountains shows much variation in landscape, but large areas consist of flat, poorly drained lowlands with marshes and bogs. The Siberian coast is generally flat and includes the deltas of many large, north-flowing rivers. Ice-covered mountains are characteristic of the Russian peninsulas of Taimyr and Chukotka. In eastern Siberia, there are several mountain ranges (e.g. Verkhoyansk, Chersky and Momsky) with peaks reaching heights of over 2500 m.

Resource 9E Arctic landscape diversity

ISBN: 9780170425285

The largest land mass in the Arctic region is Greenland. Greenland is sometimes regarded as the largest island in the world but in reality it is comprises numerous mountainous islands covered with a permanent ice cap, with central Greenland up to 3000 metres thick. Along coastal Greenland, large glaciers flow from the ice sheet into the sea where they constantly break off forming icebergs.

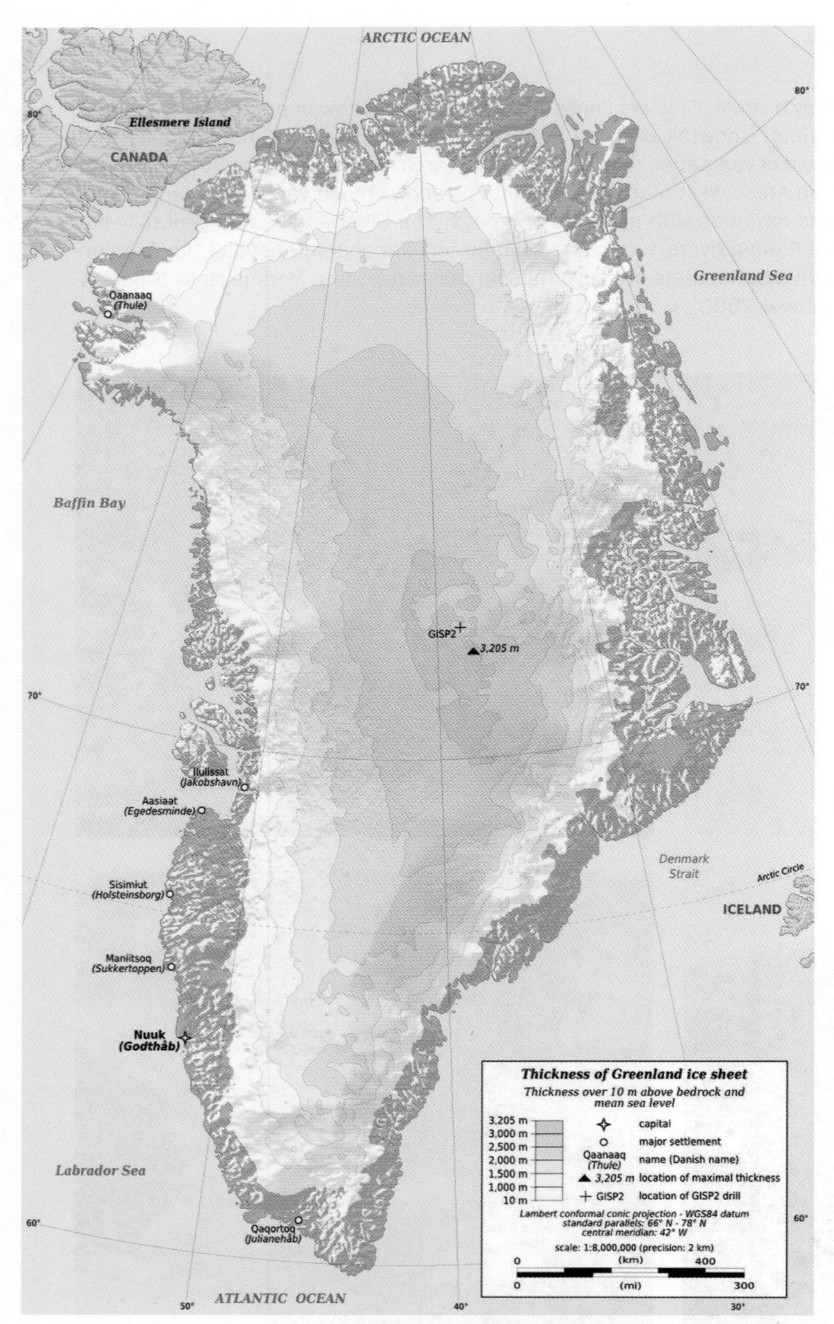

Resource 9F Greenland ice sheet thickness

Permafrost may consist of soil, bedrock or organic material and can form with or without the presence of water. The distribution of permafrost is determined by climate, particularly air temperature at ground level. In general, the soils of the Arctic are either poorly drained and underlain by ice-rich permafrost or well drained and situated over dry permafrost.

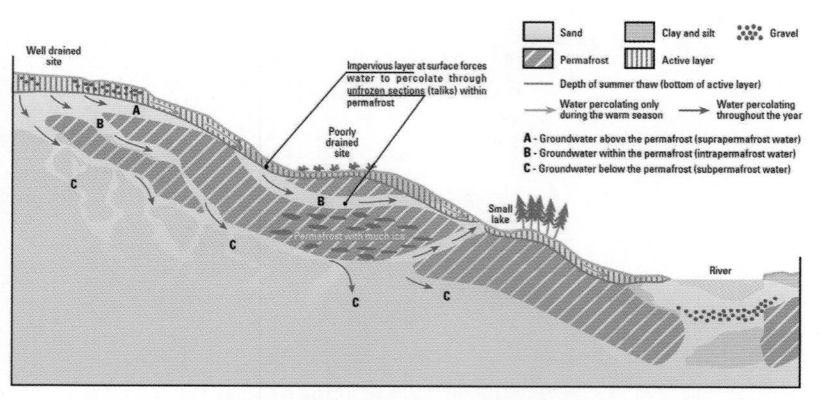

Resource 9G Occurrence of groundwater in permafrost areas

ISBN: 9780170425285

Sea ice coverage in 1980 (bottom) and 2012 (top). Multi-year ice is shown in bright white, while average sea ice cover is shown in light blue to milky white.

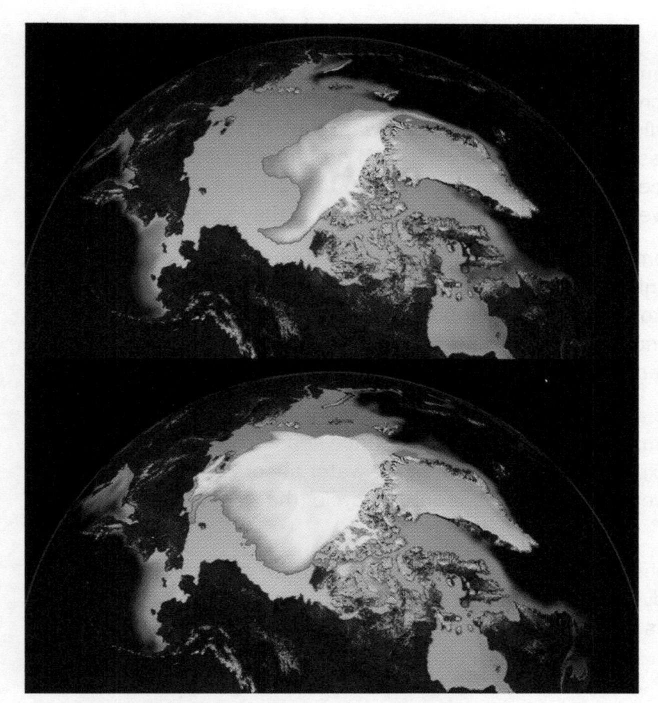

Resource 9H Sea ice coverage 1980 (bottom) and 2012 (top)

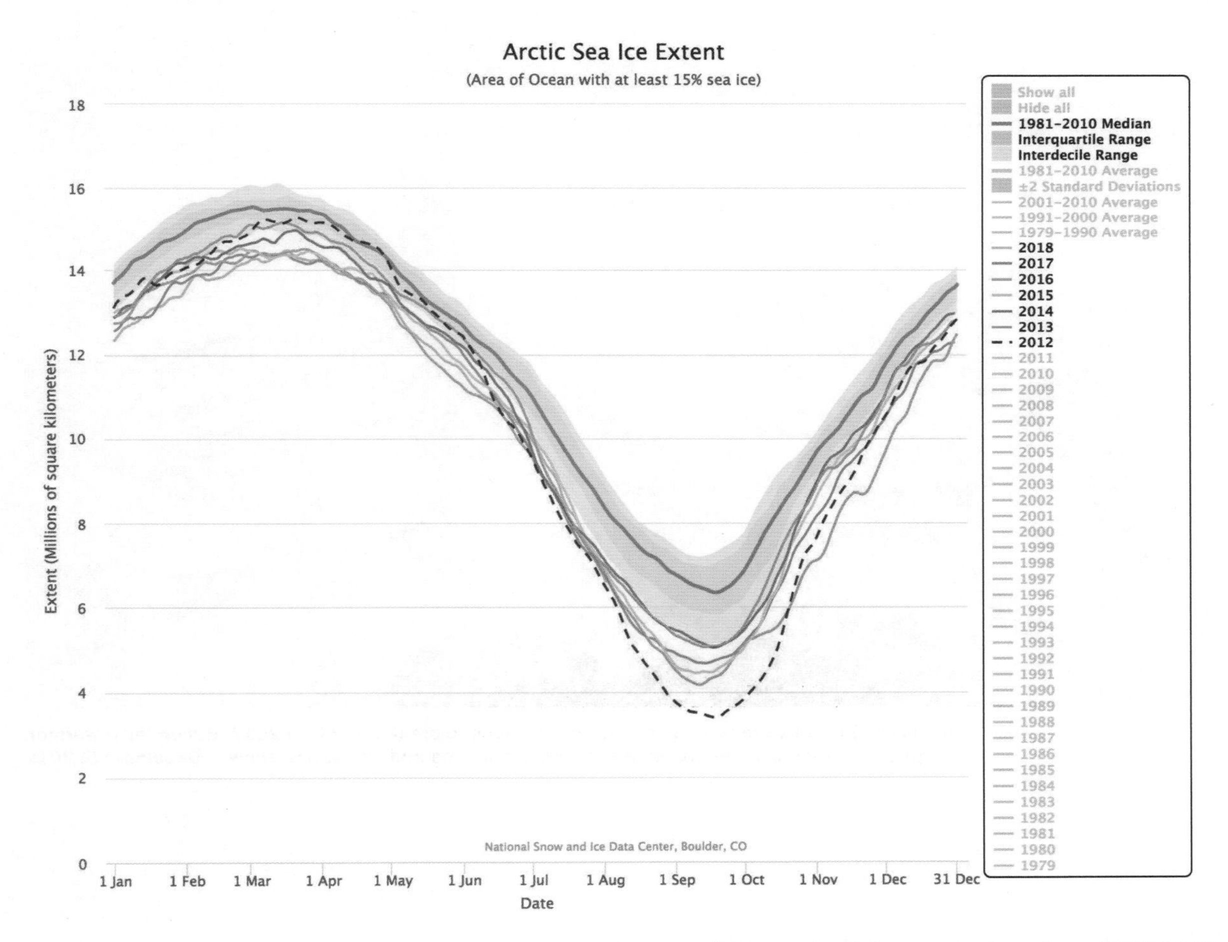

Resource 9I Seasonal variations on sea ice volume (2001-2012)

The graph above shows Arctic sea ice extent as of January 2, 2018, along with daily ice extent data for five previous years. 2018 is shown in orange, 2017 in blue, 2016 in pink, 2015 in yellow, 2014 green, 2013 in red, and 2012 in dotted green. Credit: *National Snow and Ice Data Centre*

ISBN: 9780170425285

Sea ice has a profound influence on the polar physical environment, including ocean circulation, weather, and regional climate. As ice crystals form, they expel salt, which increases the salinity of the underlying ocean waters. This cold, salty water is dense, and it can sink deep to the ocean floor, where it flows back toward the equator. The sea ice layer also restricts wind and wave action near coastlines, lessening coastal erosion and protecting ice shelves. And sea ice creates an insulating cap across the ocean surface, which reduces evaporation and prevents heat loss to the atmosphere from the ocean surface. As a result, ice-covered areas are colder and drier than they would be without ice.

Sea ice also has a fundamental role in polar ecosystems. When sea ice melts in the summer, it releases nutrients into the water, which stimulate the growth of phytoplankton, which are the base of the marine food web. As the ice melts, it exposes ocean water to sunlight, spurring photosynthesis in phytoplankton. When ice freezes, the underlying water gets saltier and sinks, mixing the water column and bringing nutrients to the surface. The ice itself is habitat for animals such as seals, Arctic foxes, polar bears, and penguins.

Sea ice's influence on the Earth is not just regional; it's global. The white surface of sea ice reflects far more sunlight back to space than ocean water does. (In scientific terms, ice has a high albedo.) Once sea ice begins to melt, a self-reinforcing cycle often begins. As more ice melts and exposes more dark water, the water absorbs more sunlight. The sun-warmed water then melts more ice. Over several years, this positive feedback cycle (the "ice-albedo feedback") can influence global climate.

Sea ice plays many important roles in the Earth system, but influencing sea level is not one of them. Because it is already floating on the ocean surface, sea ice is already displacing its own weight. Melting sea ice won't raise ocean level any more than melting ice cubes will cause a glass of iced tea to overflow.

Resource 9J NOAA Press release: 'Arctic continues to break records in 2012: Becoming a warmer, greener region with record losses of summer sea ice and late spring snow' – December 5, 2012

ISBN: 9780170425285

Turbines in Turitea

New Zealand's demand for electricity has been growing steadily and is predicted to continue to grow. To meet this growing demand, New Zealand must find new innovative ways to deliver clean, secure and affordable energy while upholding the values associated with the sustainable use of the environment.

In 2009, Mighty River Power Limited, a New Zealand electricity generation and retailing company, submitted a proposal to construct, maintain and operate a wind farm in the Turitea Reserve on farmland near Palmerston North. Following rigorous public consultation, approval was finally given for 60 wind turbines to be constructed with a maximum generating capacity of 180 megawatts.

Q1 Balancing national supply and demand

Learning Activities

Applying a geographic concept: Change

Change involves any alteration to the natural or cultural environment. Change can be spatial and/or temporal. Change is a normal process in both natural and cultural environments. It occurs at varying rates, at different times and in different places. Some changes are predictable, recurrent or cyclic, while others are unpredictable or erratic. Change can bring about further change.

Today, more than half of New Zealand's electricity is generated from hydroelectric stations, and with the addition of other energy sources such as geothermal, biomass, solar and wind, renewable energy resources account for 77 percent of New Zealand's electricity generation. Electricity consumption has increased at approximately 2.5 percent a year since the 1970s. That trend is forecast to continue growing at a rate of approximately 1.5 percent a year through to 2050.

Refer to the definition of change above and **Resources 10A – 10C** to answer this question.

a With reference to specific energy sources, describe the change in the generation of electricity.

i General trend:

______

______

______

ii Resource specific trends:

______

______

______

______

______

______

______

______

______

ISBN: 9780170425285

Learning Activities

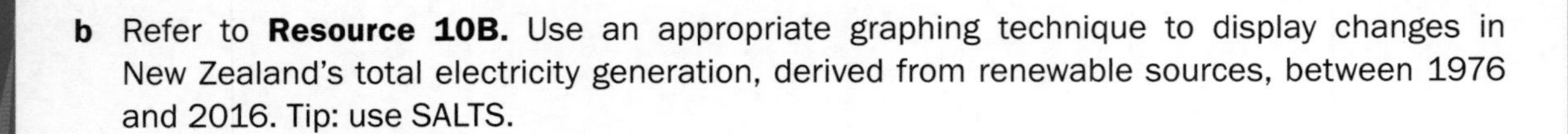

b Refer to **Resource 10B.** Use an appropriate graphing technique to display changes in New Zealand's total electricity generation, derived from renewable sources, between 1976 and 2016. Tip: use SALTS.

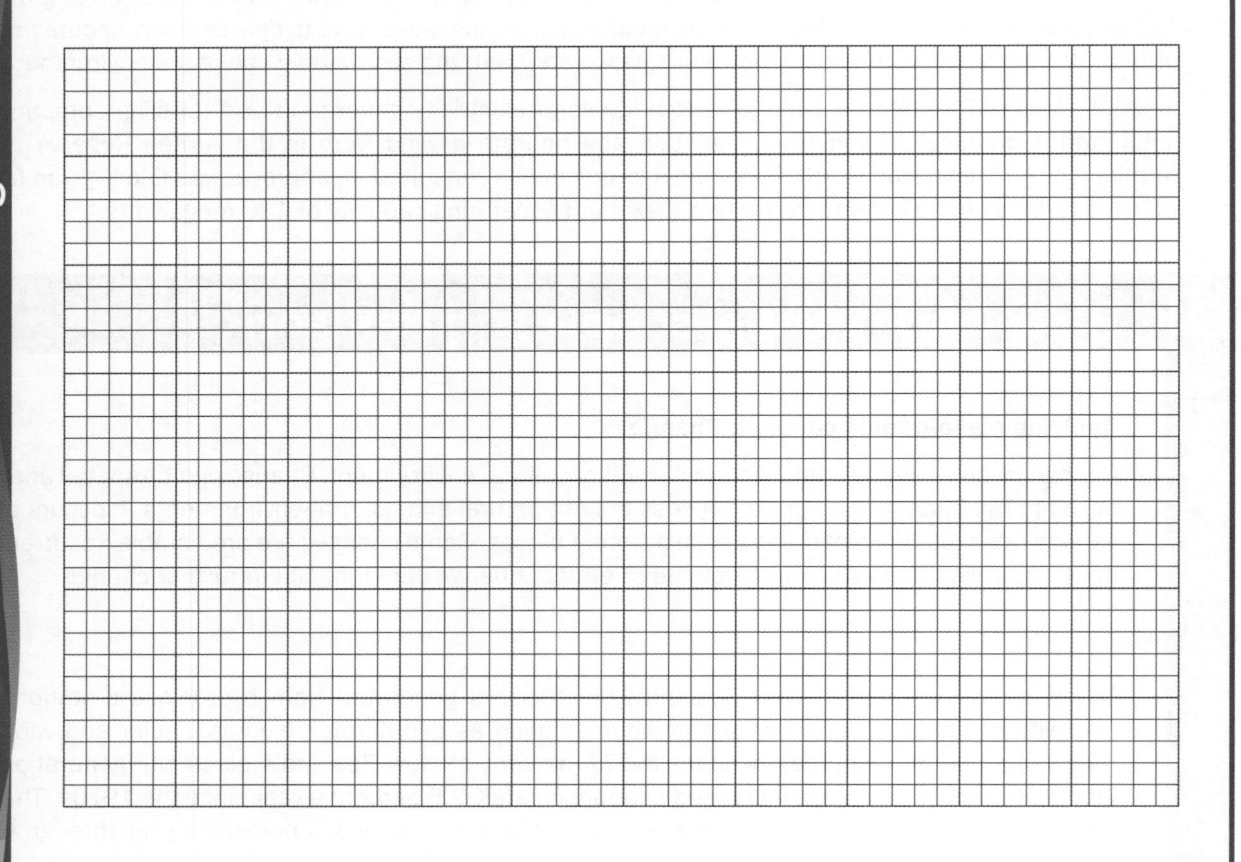

c Refer to **Resource 10C**. Suggest two reasons why the increase in the demand for electricity is likely to be slower into the future (1.5 percent per annum) than it has been in the past (2.5 percent per annum).

i Reason 1:

ii Reason 2:

ISBN: 9780170425285

Learning Activities

d Describe the general trend in electricity consumption shown in the graph.

e Analyse the change in energy consumption for the following sectors:

i Industrial

ii Residential.

ISBN: 9780170425285

Learning Activities

Q2 Turitea: site and situation

Applying a geographic concept: Environments

Environments may be natural and/or cultural. They have particular characteristics and features, which can be the result of natural and/or cultural processes. The particular characteristics of an environment may be similar to and/or different from another.

The site for Turitea is approximately 10 kilometres southeast of Palmerston North primarily along a 14-kilometre ridge in the northern Tararua Range. The wind resource at the Tararua Ranges is exceptional, due to the high average wind speeds at elevated locations.

Refer to the definition of environments above and **Resources 10D – 10G** when answering this question.

a Complete the précis map on the following page by accurately locating and labelling the following features:

i Area of the proposed Turitea Wind Farm

ii Tararua Range

iii The area of settlement (Palmerston North), north of the Manawatu River

iv Area of farmland west of the Tararua Range.

b Identify the direction the camera was facing in **Resource 10F**.

c Account for the size and extent of Turitea Wind Farm.

ISBN: 9780170425285

Learning Activities

Key

ISBN: 9780170425285

Learning Activities

d Analyse one characteristic of the natural environment and one characteristic of the cultural environment from those listed in **Resource 10G**.

i Natural environment characteristic:

ii Cultural environment characteristic:

ISBN: 9780170425285

Q3 Local versus national perspectives

Learning Activities

Applying a geographic concept: Perspectives

Perspectives are ways of seeing the world that help explain differences in decisions about, responses to, and interactions with environments. Perspectives are bodies of thought, theories or worldviews that shape people's values and have built up over time. They involve people's perceptions (how they view and interpret environments) and viewpoints (what they think) about geographic issues. Perceptions and viewpoints are influenced by people's values (deeply held beliefs about what is important or desirable).

Refer to the definition of perspectives above and **Resources 10H** and **10I** to support your answer.

a Conduct a costs/benefits analysis of the Turitea proposal comparing local perspectives with the perspectives of the nation.

	Costs	Benefits
Local perspective		

ISBN: 9780170425285

	Costs	Benefits
National perspective		

b With reference to specific information from throughout the chapter, discuss the following statement:

'The national benefits of renewable generation outweigh the adverse effects borne at the local scale.'

ISBN: 9780170425285

Learning Activities

ISBN: 9780170425285

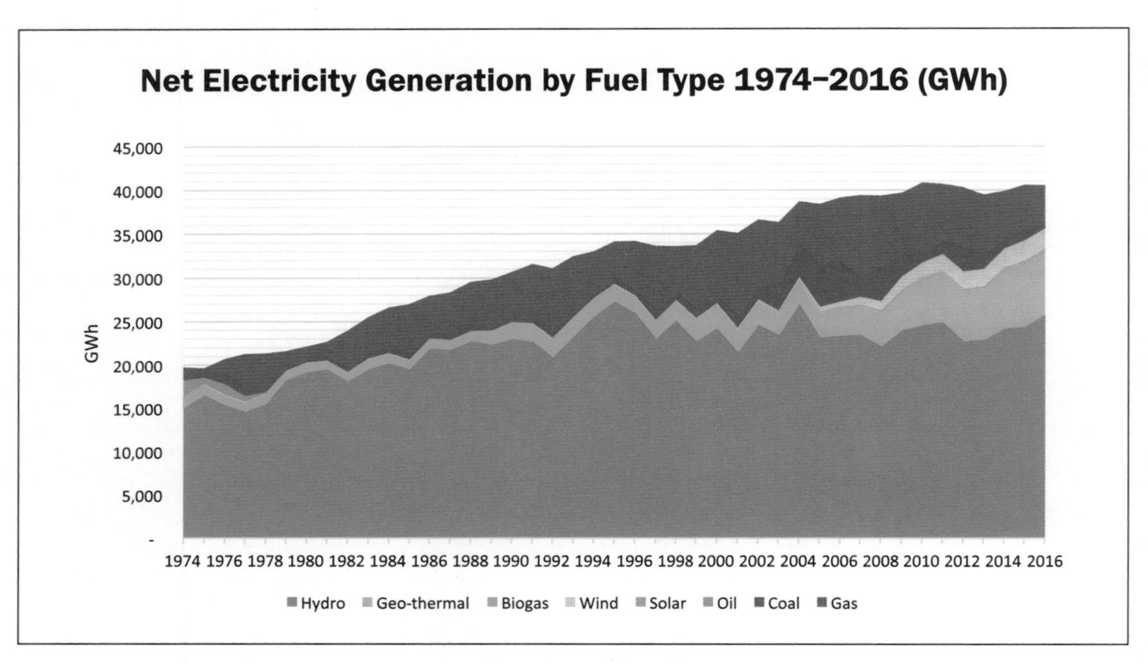

Resource 10A Net electricity generation by fuel type 1974–2016

The latest international comparison shows that New Zealand has the second-highest contribution of renewable energy to Total Primary Energy Supply (TPES) in the OECD (behind Iceland, and ahead of Norway).

Increased electricity generation from geothermal energy and reduced electricity generation from coal have driven the rapid increase in renewable energy's share of TPES in the last few years. As geothermal fluid is much lower in temperature than steam produced by a coal or gas boiler, the transformation efficiency of geothermal energy is significantly lower. The low transformation efficiency of geothermal energy (approximately 15%) contributes to New Zealand's relatively high renewable TPES when compared with most other countries.

Although geothermal energy's share has increased rapidly in the last three years, oil continues to dominate New Zealand's TPES. (Ministry of Economic Development)

Year	Renewable				Non-renewable					TOTAL	% Renewable
	Hydro	Geothermal	Wind	Sub-total	Coal	Oil	Gas	Waste Heat	Sub-total		
1976	55.79	39.96	26.59	122.34	60.90	179.12	37.54	1.28	278.84	401.18	30%
1981	70.84	37.76	29.64	138.24	48.88	160.56	43.64	1.67	254.75	392.99	35%
1986	79.54	40.75	29.70	149.99	52.40	156.11	167.90	1.61	378.02	528.01	28%
1991	82.41	64.87	41.03	188.31	46.71	184.10	195.80	1.61	428.22	616.53	31%
1996	94.25	63.46	44.80	202.51	39.49	225.21	203.32	1.62	469.64	672.15	30%
2001	78.04	75.93	59.87	213.84	59.50	250.37	247.48	1.21	557.94	771.78	28%
2006	84.86	86.35	71.51	242.43	85.41	281.82	154.03	1.83	523.17	765.60	32%
2011	90.28	159.05	72.60	321.57	60.74	276.21	158.68	1.52	498.65	820.22	39%
2016	93.26	201.84	69.98	365.08	52.09	293.27	195.82	1.41	542.59	907.68	40.27

Resource 10B Total Primary Energy Supply (TPES)

ISBN: 9780170425285

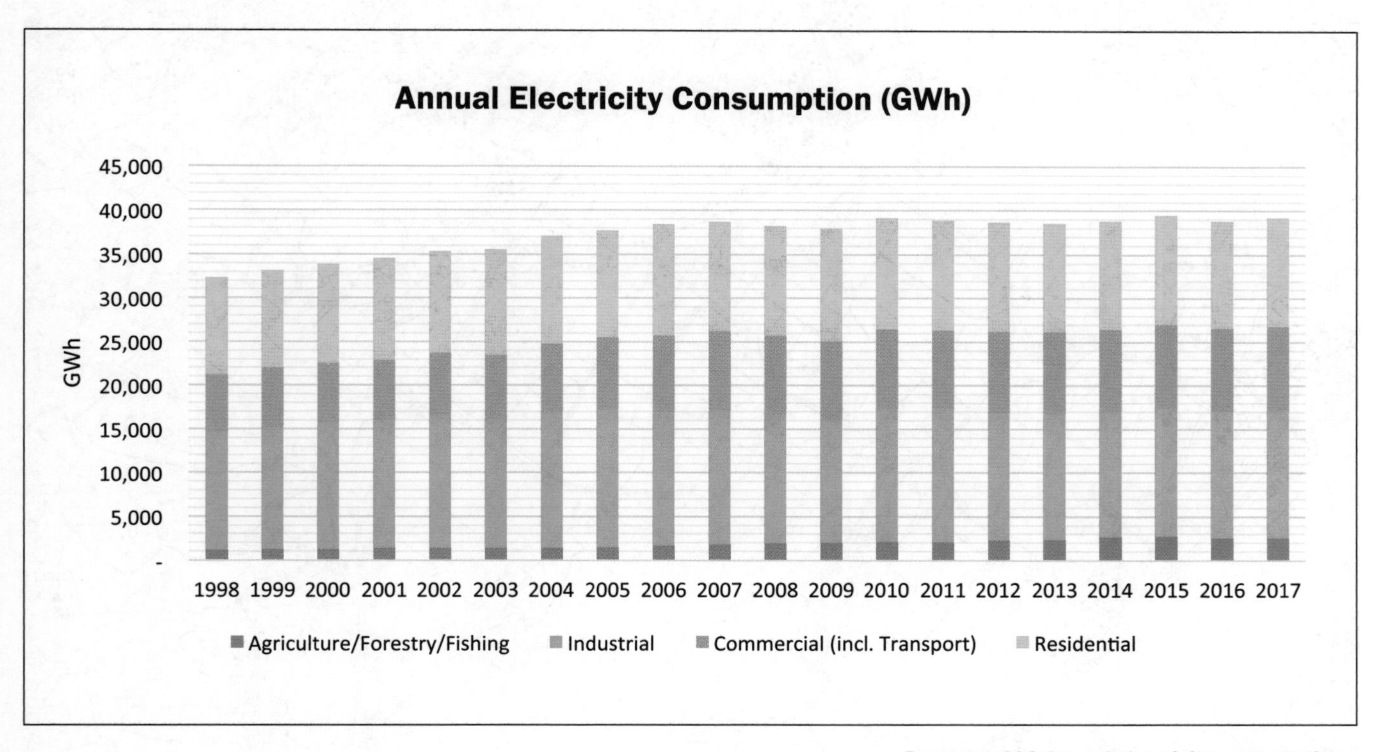

Resource 10C Annual electricity consumption

ISBN: 9780170425285

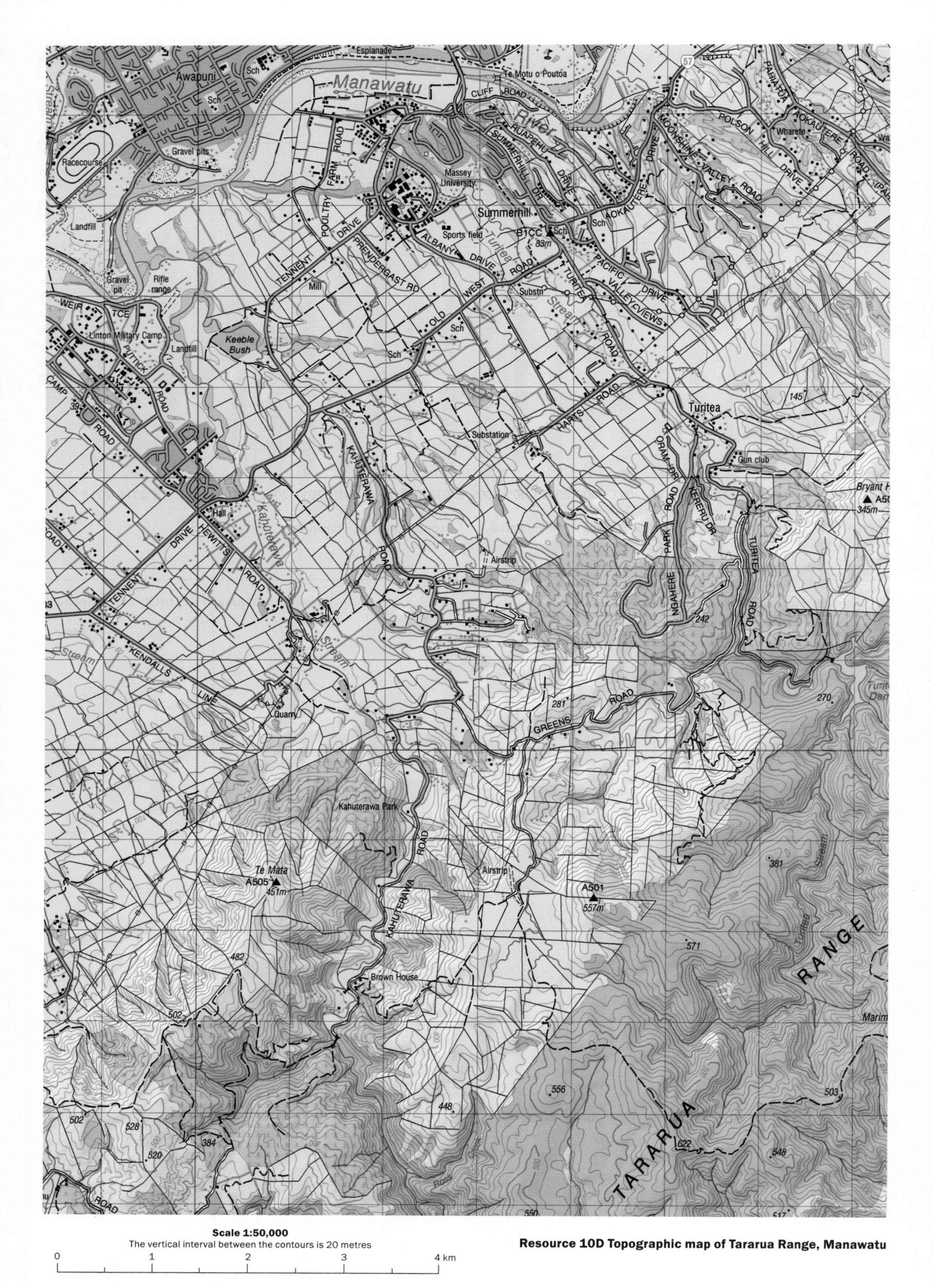

Scale 1:50,000
The vertical interval between the contours is 20 metres

0 1 2 3 4 km

Resource 10D Topographic map of Tararua Range, Manawatu

ISBN: 9780170425285

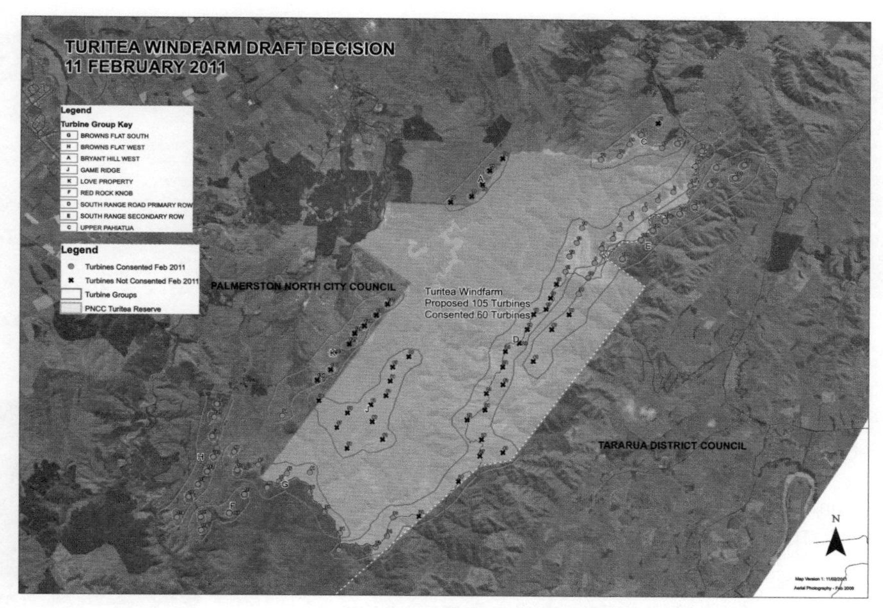

Resource 10E Size and extent of the Turitea Wind Farm

Resource 10F Current Palmerston North cityscape and Tararua Range

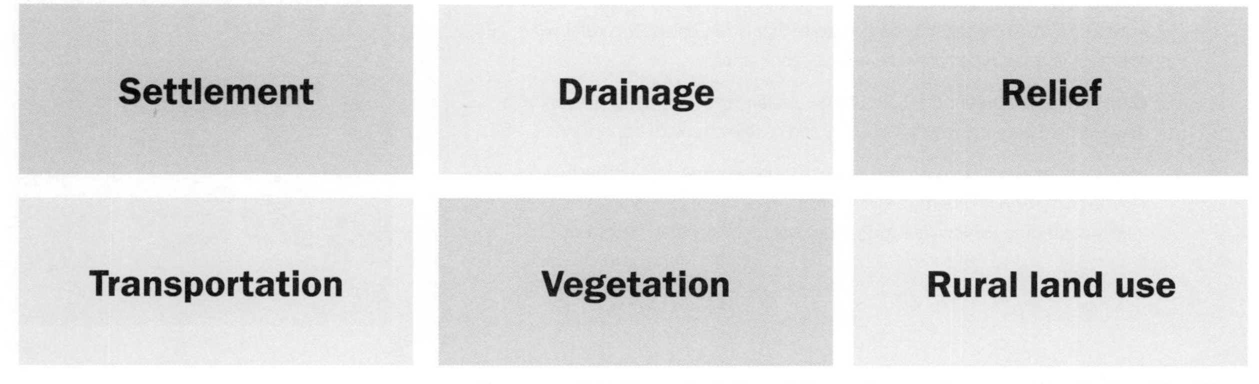

Resource 10G Characteristics of the environment surrounding Turitea Wind Farm

ISBN: 9780170425285

www.windenergy.org.nz

Benefits of wind farms

Wind farms create a range of economic, social and environmental benefits at global, national and local levels. This fact sheet highlights some of those benefits.

A series of case studies about some of the benefits of wind farms are available at: http://windenergy.org.nz/resources/resources/case-studies. These case studies discuss the benefits of wind farms in terms of Business and Community Opportunities, Improving Electricity Supply, Farming the Wind; and Local Economic Benefits.

Global benefit

Modern societies rely on access to electricity. Wind energy is recognised globally as one of the most sustainable forms of electricity generation. This is largely because the use of wind energy, a renewable natural resource, slows the consumption of finite and exclusive fuels, which preserves the natural environment and thus provides for future generations.

National benefits

Increasing New Zealand's electricity supply from wind energy will put New Zealand on a better path towards sustainable development:

- Wind farms generate electricity using an infinite and abundant renewable energy resource and therefore wind farms help reduce reliance on finite fossil fuels.
- Electricity generated from the wind creates no more greenhouse gas emissions than hydro-electricity, and much less than all other forms of electricity generation available in New Zealand, including geothermal and solar.
- As a clean, safe and green form of electricity generation, wind energy helps promote New Zealand's 100% Pure brand.
- Wind energy is New Zealand's preferred form of new electricity generation. Studies commissioned by the New Zealand Energy Efficiency and Conversation Authority (EECA) consistently demonstrate that New Zealanders prefer wind energy over other types of new electricity generation.
- Wind farm development throughout New Zealand provides national economic development benefits through industry and business growth opportunities.
- Wind farms are long-term infrastructure that utilise a free and reliable fuel supply. This helps make wind energy a low risk form of electricity generation that can readily contribute to New Zealand's wider electricity generation system.

New Zealand's National Policy Statement for Renewable Electricity Generation identifies that renewable electricity generation activities have national, regional and local benefits including but not limited to:

a) maintaining or increasing electricity generation capacity while avoiding, reducing or displacing greenhouse gas emissions;

b) maintaining or increasing security of electricity supply at local, regional and national levels by diversifying the type and/or location of electricity generation;

c) using renewable natural resources rather than finite resources;

d) the reversibility of the adverse effects on the environment of some renewable electricity generation technologies;

e) avoiding reliance on imported fuels for the purposes of generating electricity.

ISBN: 9780170425285

National benefits continued...

- Wind farms can be small or large. This means they can be located and designed to suit a wide variety of needs, from a small amount of generation feeding into a rural network to a large wind farm powering a city.
- Wind energy is a competitive price taker in the electricity market. It helps reduce the spot price of electricity.
- The environmental effects of wind farms are the most reversible of any form of utility-scale electricity generation in New Zealand. Wind farm infrastructure can be decommissioned and taken off site leaving only the physical footprint of land modification resulting from internal roads and foundations.
- Wind farms can secure the value of the existing land, which in turn can protect natural values for future generations by preventing intensive land development. In some cases wind farms can help improve New Zealand's overall biodiversity values through supporting ecological restoration.
- Wind farms contribute to New Zealand's national and international obligations for reducing greenhouse gas emissions and addressing climate change.

Local benefits

- Most wind farms are built on rural land. Wind farms provide landowners with a new income stream that can improve the return from their land and allow valued agricultural business to continue.
- Wind farm construction creates significant local economic benefits through local spend, opportunities for business development including tourism, community projects and research. Wind farm construction is often the largest capital works project in rural areas.
- Wind farms can be scaled to fit demand, resource availability and site constraints. Therefore wind farms can potentially operate in each region of New Zealand.
- Wind farms can increase the stability of the local electricity network and increase security of supply at a local level thus contributing to a region's self-sufficiency.
- Wind farms can enhance recreation values. In some cases wind farm development can provide opportunities for improving public access to previously inaccessible areas of high recreational value.
- Wind farms can enhance heritage values. In some cases wind farm development can provide opportunities for improving a range of heritage engagement enhancement opportunities, for example, improving public access to selected historic sites and areas.
- Wind farms can enhance the local road network. Developers will often upgrade local roads to enable large wind turbine components to be transported to site.
- Wind farms can enhance local biodiversity values. In some cases wind farm developers can contribute to the restoration or improvement of local biodiversity values, including protecting indigenous biodiversity on farm land.
- Wind farms can enhance amenity values. Many people like the look of wind farms, although in every case there is a degree of subjective judgement involved. In New Zealand, communities have used wind turbines to promote their wellbeing and 'sense of place'.

Gebbies Pass

More Information

Find out more about wind energy and wind farms in New Zealand at www.windenergy.org.nz.

NZ Wind Energy Association

PO Box 553, Wellington 6140,
New Zealand

The New Zealand Wind Energy Association (NZWEA) is an industry association that works towards the development of wind as a reliable, sustainable, clean and commercially viable energy source. We aim to fairly represent wind energy to the public, government and the energy sector. Our members include 80 companies involved in New Zealand's wind energy sector, including electricity generators, wind farm developers, lines companies, turbine manufacturers, consulting firms, researchers and law firms.

JUNE 2013

Resource 10H Wind energy

ISBN: 9780170425285

Turitea Wind Farm supporters put their side

After listening to days of argument against the planned Turitea Wind Farm near Palmerston North, supporters of the project have now had their say.

Farmer Kathy Love yesterday told the board of inquiry considering the proposal that the wind farm would be a "pretty benign development".

She described the Tararua Ranges as a "big, raw, energetic landscape".

"I think it can take turbines quite happily," she said.

A hearing on the Turitea proposal wrapped up for the year yesterday, after seven weeks of expert evidence and public submissions.

It is likely to resume about March, by which time Mighty River Power will have completed a redesign of the farm's layout.

Mrs Love, who would receive a rental from Mighty River Power for a series of turbines on her property, said the landscape was already industrial in that it was used for New Zealand's biggest industry – agriculture.

The farm land was "marginal". Turbines could reduce the need for subdivisions if farming was to become unviable, she said.

Landowner Joseph Poff said most Palmerston North people supported the project – they had not been swayed by a negative campaign run by a few people, he said.

Mr Poff, whose submission was accompanied by what sounded like the jingling of coins from the public gallery, said the proposal was "the biggest no-brainer in the history of Palmerston North".

Mr Poff said he could sometimes hear nearby Te Rere Hau turbines from his property, but annoyance depended on "whether you choose to be bothered by that sound".

Renewable energy from wind farms was part of the solution for climate-change related problems, he said.

Mrs Love said the project attracted antagonism because people feared change.

Failing to use the wind resource would be a "criminal waste".

Turitea Reserve would also benefit from money generated by the project – this could be used for better pest control, she said.

Husband John Love said the couple had a strong environmental record – retiring 75 hectares of native bush and contributing to fencing to keep stock out of the reserve.

A series of submitters yesterday said it would be wrong to allow the "industrialisation" of a nature reserve. Palmerston North resident Brent Barrett suggested the meaning of "reserve" had changed to: "If you rock up with some cash, we'll be happy to see you."

Tararua wind farms 'a cancer'

Wind turbines are like "a cancer" spreading along the Tararua Ranges, Palmerston North city councillor Michael Feyen has told a board of inquiry.

Yesterday he called for the rejection of Mighty River Power's Turitea Wind Farm proposal, arguing the public had not been adequately informed. He opposed wind-farm development south of the Pahiatua Track because "the ranges, to me, were looking like a wind turbine junk yard".

Mr Feyen, hurried along by the board about a dozen times, also said the city's drinking water was at risk from sediment caused by the project.

Two other city councillors, Chris Teo-Sherrell and Bruce Wilson, presented their views on Tuesday.

Mr Teo-Sherrell detailed to the board his efforts to cut his personal carbon footprint, before arguing the Turitea proposal should not go ahead. Climate change was not yet so serious that it warranted degrading "special places". Residents near the proposed farm should not be unwilling participants in an experiment, he said.

Mr Wilson was worried about the cumulative visual effect of "yet another wind farm". He conceded the Tararua Ranges were not as outstanding as other landscapes in New Zealand, but they were "our outstanding landscape".

Mr Wilson said that he was "ashamed" of various turbine styles and sizes on the Tararua and Ruahine Ranges.

Palmerston North could be the "sacrificial lamb" that prompted councils throughout the country to tighten up the guidance issued in district plans, he said.

Resource 10I Wind farm supporters put their side

ISBN: 9780170425285

Answers

From Recovery to Rebuild

1a Christchurch is located east of the Southern Alps on the Canterbury Plains. The Plains comprise young sediments (gravel, sand, mud) eroded from the Southern Alps and transported by the Rakaia and Waimakariri rivers.

1b Responses to this question must consider aspects of both the natural and cultural environment of Christchurch and Canterbury. For example: Christchurch is located 200 km east of an active tectonic plate boundary (Alpine Fault). East of the Alpine Fault a number of smaller fault lines exist beneath the Canterbury Plains. This combined with urban Christchurch's relatively high population density poses a threat to the people of Christchurch.

2a i The most suitable graphing technique for this question is either a flow-line map or a proportional symbol map. Note: This graphical map should only show the movement of people away from Christchurch after the 2011 earthquake.

ii Approximately four-fifths (81.4%) of people who moved away from Christchurch after the 2011 earthquake moved to other parts of the Canterbury region. The Auckland region (4.1%) and Otago region (3.3%) were significant flows. Population flows to other regions are negligible (< 1%).

2b i Christchurch experienced more fatalities (185) than the larger Loma Prieta (65) and Northridge (60) earthquakes but much fewer than that of the Kobe earthquake. Fewer building units (6000) were damaged in the Christchurch earthquake compared with comparable events. The wholesale destruction of the Christchurch CBD and destruction of the major port at Kobe suggests that Kobe possibly provides the closest model to the Canterbury event. This would suggest that expectations of recovery time are likely to be closer to the 10+ years for Christchurch. Note: Hurricane Katrina should not be considered a comparative event.

ii Teacher to mark this question. Answers should comment on net migration being greater for females relative to males in the weeks/months following the February 2011 earthquake (Resource 1E).

3a i The Christchurch recovery process will encourage inward migration to Christchurch as people deem it safe to return or in response to new employment opportunities.

ii The Christchurch rebuild will create a demand for skilled labour in occupations and industries directly related to the reconstruction. Demand for labour skills are likely to include builders, stonemasons, engineers (Resource 1H).

3b Teacher to mark this question. Responses must acknowledge the three phases/stages of recovery: restoration (immediate), reconstruction (short term) and improvement (medium to longer term).

The New Fossil Fuel Frontiers

1a i 33°S to 37°S, 177°E to 178°W

ii The basin extends approximately 300 km with a northeast to southwest orientation and extends 10 km west to east. Its approximate area is 30,000 km^2 (Resource 2B).

1b Teacher to mark this question. Responses to this question should reference specific geological periods: Cretaceous, Paleogene and Neogene.

2a Positive impacts: increased access to jobs, tax and royalty income, and regional development. Negative impacts: seismic testing and drilling may impede customary rights such as fishing and seafood collection; diminishing integrity of the marine environment may lead to diminished access to resources which may in turn lead to local iwi's loss of cultural identity (Resource 2D).

2b Only potential effects on the natural environment should be discussed in responding to this question. Potential effects based on the experience of recent oil pollution events may include: impacts on marine species (fisheries and seafood) including birds (Resource 2F), e.g. blue penguins, pied shags and the rare New Zealand dotterel (Resource 2G), and coastal environments, e.g. wetlands, beaches.

3a i Teacher to mark this question.

ii Teacher to mark this question. Responses to this question must give valid reasons or examples to support the discussion.

Shanghai Sprawl

1a 31°15' N, 121°26' E (or 31.2000° N, 121.5000° E)

1b Teacher to mark this question.

1c South-southeast (SSE) to southeast (SE)

1d i Shanghai's location on the east coast facilitates regional and international trade via its large seaport (Resource 3D).

ii Low-lying land is susceptible to flooding and subsidence (Resource 3E).

iii Any discussion that is supported with evidence from the resources is acceptable.

2a i The most suitable graphing technique for this question is a flow-line map or proportional symbol map. A flow-line map must indicate the direction and flow of migrants to Shanghai.

ii The pattern of internal migration is generally from centrally located provinces (Anhui, Jiangsu, Henan and Sichuan) in southern China, to Shanghai in the east.

2b i

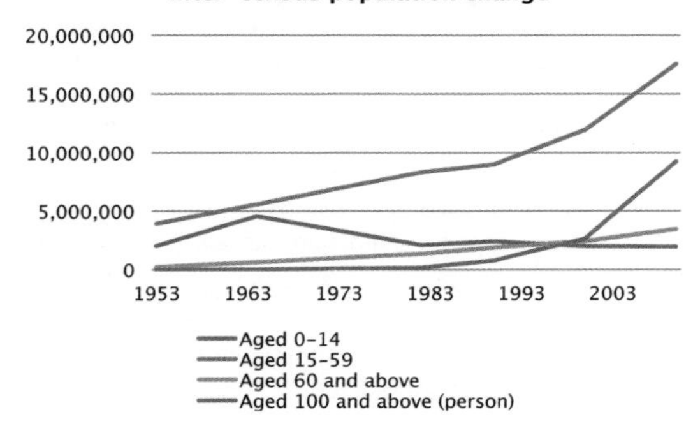

ISBN: 9780170425285

ii Shanghai's population is ageing. This is evidenced by the decrease in the number of inhabitants aged 0–14 and the increase in inhabitants aged 100 and above. Between 1953 and 2013, the greatest increase came from the working age groups (15–59). Detailed answers will include some quantification of the data contained in Resource 3F.

iii Shanghai's population pyramid is diamond shaped. It has a narrow top and narrow base. The central portion of the pyramid represents the working age group, which has grown significantly since 1953 (Resource 3F).

3a Teacher to mark this question. Identified problems should include: water shortages and issues relating to building on marginal land, e.g. subsidence. Other possibilities include: urban sprawl, housing shortages and overdevelopment.

Nuclear Meltdown

1a Responses to this question must support the case that the Chernobyl nuclear disaster was human induced (i.e. a technological disaster) and not the result of natural processes. Responses should be supported with detailed evidence from Resource 4D.

1b Teacher to mark this question. Cycle should begin with the 9-magnitude earthquake that struck Japan northeast of Tokyo on 11 March 2011 (Resource 4E).

1c Responses to this question should recognise the contrasting causal factors that differentiate the two events, i.e. direct human error led to the Chernobyl disaster whilst the underlying causal factor leading to the Fukushima Daiichi disaster was a sequence of two extreme natural events (an earthquake followed by a devastating tsunami) (Resources 4D and 4E).

2a 51°16' N, 30°14' E (or 51.3896° N, 30.0991° E)

2b Approximately 1500–1900 km^2 (Resource 4G).

2c i Using PQE, responses to this question should identify that at a reading of 40 Ci/km^2, the highest concentration of surface ground contamination of caesium-137, occurred in the area immediately surrounding Chernobyl, Ukraine (part of the former Soviet Union), and the region directly to the north. Transboundary contamination of radioactive material (caesium-137) also dispersed beyond the location of Chernobyl, Ukraine. For example, Suomi (Finland), Sverige (Sweden) and Osterreich (Austria) also experienced ground contamination of up to 5 Ci/km^2 (Resource 4F).

ii Using PQE, responses should identify that the highest deposition of caesium-137 (6,400nSv/h) was localised within a 25 miles radius of Fukushima Daiichi power plant, with high concentrations also occurring in the area to the north west of the power plant as far as Fukushima City. In contrast, the large region to the south west of Fukushima Daiichi received lower levels of surface ground contamination (less than 100nSv/h) (Resource 4G).

3a i Teacher to mark this question. Responses to this question should only evaluate aspects of the natural environment (e.g. natural habitats, air, soil and water).

ii Teacher to mark this question. Responses to this question should only evaluate aspects of the cultural environment (e.g. food contamination and the potential effects of radiation on the human body).

Destination Fiordland

1a Teacher to mark this question. Ensure all mapping conventions are satisfied (i.e. title, use of key, north point).

1b 1:50,000

1c Teacher to mark this question. A consistent scale is required on both axes. The cross-section should be accurately labelled and/or annotated with appropriate landscape features (relief, drainage, vegetation).

1d i Acceptable answers to this question include: its perceived aesthetic beauty of its unique and relatively untouched natural environment; its World Heritage status.

ii Acceptable answers to this question could include: limited accessibility in terms of distance and travel times from nearby settlements; the effect of distance decay; limited accommodation facilities within the national park.

2a i Teacher to mark this question.

ii Teacher to mark this question. Perspectives discussed must correlate with the placement of viewpoints on the continuum in **i**. Responses must identify valid reasons or evidence drawn from specified resources.

3a Teacher to mark this question. Responses to this question must demonstrate an understanding of the geographic concept of sustainability as applied to Riverstone Holdings' monorail proposal. The comprehensive analysis of the environmental sustainability of the proposal needs to draw from the issues specified in a range of resources, particularly Resource 5F, which refers to the DOC requirement for Riverstone Holdings to develop and enforce a range of environmental management plans (including a Vegetation and Habitat Management Plan) to ensure the environmental impacts of the construction and ongoing operation of the monorail are appropriately monitored and managed.

City Rail Link

1a i Most new arrivals to Auckland are from overseas. There has also been a net positive flow of people moving to Auckland from other regions of New Zealand. However, since 1996–2001, this latter trend has reversed (Resource 6A).

ii PQE: Resource 6A shows that the population growth rate of Auckland is greater than that of the rest of New Zealand in all time frames. For example, Auckland annual population growth over 10 years was 2% compared to 1.2% for all of New Zealand for the same period (Resource 6A).

1b i Inter-census growth between 2006 and 2013 led to an increase in population density in almost all areas of Auckland. Most significant changes occurred on the urban-rural fringe of South Auckland and the CBD. Responses to this question should include some reference or quantification of the data contained in Resource 6B.

ii Inter-census growth between 2006 and 2013 was slower than the previous inter-census growth rate (8.5% versus 12.5%). Responses to this question should include some reference or quantification of the data contained in Resource 6B.

2a i Teacher to mark this question. The most appropriate graphing technique to display mobility is the flow-line map constructed using data from Resource 6C. Ensure all mapping conventions are satisfied (i.e. title, use of key, north point).

ISBN: 9780170425285

ii Teacher to mark this question. Detailed responses will include quantification of movements and reference to the five places of residence listed in Resource 6C.

2b 1.42%

2c

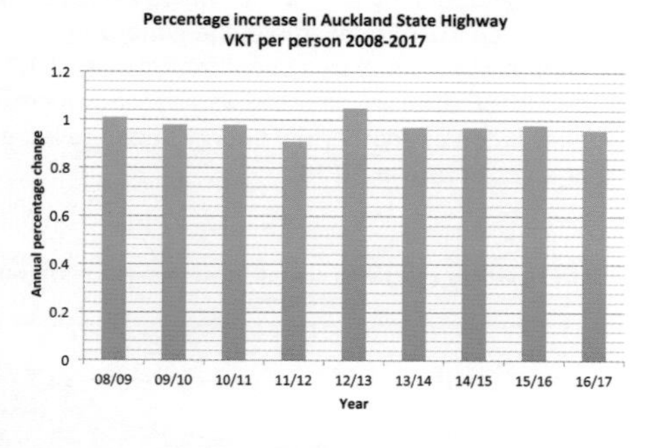

2d Responses should draw from the ideas presented in Resource 6E.

3a Teacher to mark this question. Responses to this question must demonstrate an understanding of the geographic concept of sustainability as applied to the City Rail Link. The evaluation of the environmental sustainability of the proposal should appraise the strength and weaknesses of the project as specified in a range of resources.

Transmission Gully

1a Teacher to mark this question. Ensure all mapping conventions are satisfied (i.e. title, use of key, north point).

1b 33 km

1c The natural environment consists of V-shaped valleys and rolling hill country and contains a number of freshwater streams. Responses to this question must not refer to cultural features of the environment such as farmland or transmission lines.

1d From SH1 at MacKays Crossing, north of Paekakariki, the route rises steeply to the Wainui Saddle and follows Transmission Gully down to Porirua Harbour's Pauatahanui Inlet. It continues south around the outer edge of the Porirua urban area, at one point crossing a 300-metre-long and 90-metre-high bridge, and rejoins SH1 at the boundary of Porirua and Tawa. The length is 26 km, with a maximum gradient of about 8.3% (Resource 7D).

1e The new route is safer for commuters; it offers a congestion-free option between Wellington and outlying settlements making the journey to work quicker and therefore more fuel-efficient; conversely the new road reduced congestion around coastal communities (e.g. Porirua and Pukerua Bay).

2a Responses need to identify ways in which the project will benefit the whole country. Ideas may relate to matters of safety and the associated costs of dealing with congestion and traffic accidents; and/or discuss how the project will benefit the wider economy.

2b

	Positive	Negative
Social	Reduced travel times between Wellington and outlying settlements. Fewer traffic accidents/ fatalities.	Potential for traffic bottlenecks to appear in new locations/ intersections servicing the new motorway.
Economic	Increased transport efficiencies relating to the movement of freight in and out of the region.	The local businesses on the former coastal route may suffer economically; loss of some farmland.
Environmental	Preservation of coastal environments susceptible to erosion.	Risk of increased sedimentation into freshwater streams; potential impact on stream ecology; significant re-contouring of the landscape required; modification of 10 km of streams; loss of 10 ha of native forest.

3a Teacher to mark this question. Responses to this question must demonstrate an understanding of the geographic concept of sustainability as applied to the Transmission Gully project. The evaluation of the environmental sustainability of the proposal should appraise the strength and weaknesses of the project as specified in a range of resources.

Rwanda: Land of a Thousand Hills

1 Teacher to mark this question.

2a i Rank 1: Africa
ii Rank 2: Europe
iii Rank 3: America

2b Option A:

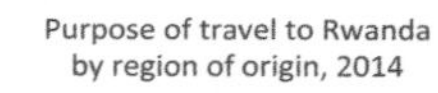

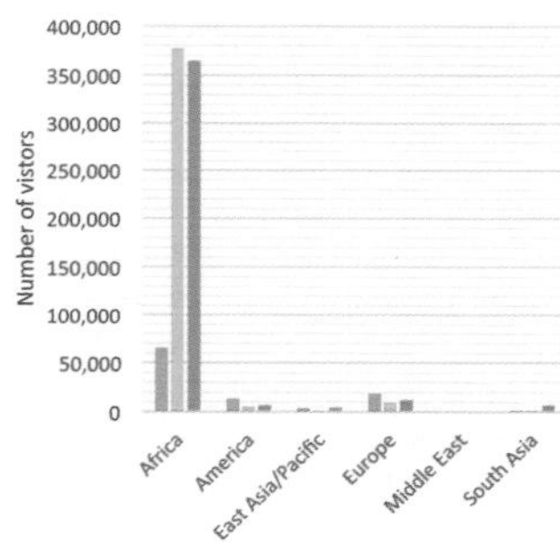

Option B:

Purpose of travel to Rwanda by region of origin from, 2014

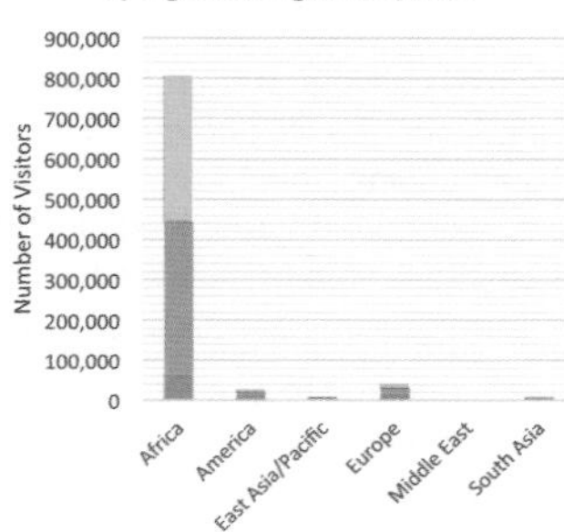

ISBN: 9780170425285

2c Tip: use PQE: For 2014, the majority of African visitors to Rwanda came for business, conferences or other work-related reasons (364,690 visitors) whereas the primary purpose of travellers visiting Rwanda from other regions was to have a holiday or vacation. The exception to this was visitors from the Middle East whose primary purpose for visiting Rwanda was for business, conferences or other work-related reasons (711 visitors) closely followed by visiting friends and family (587 visitors).

2d Tip: use PQE: Although there was a decline in the contribution of tourism to total GDP from 2008 to 2010, the contribution of tourism to GDP has steadily grown since 2011 with only a recent plateau in growth occurring since then. For example, the growth of tourism contribution to GDP reached a low of -11.5% in 2009 but later expanded to a high of 15% in 2012.

3a One social consequence: Development of tourism-related organisations, tourism has contributed to rehabilitation of towns and communities. Tourism encourages political stability. Domestic tourism has also increased.
One economic consequence: New jobs have been created in the area of tourism (e.g. gorilla trackers, porters and park rangers) and conservation. Tourism brings $400 million into the country.
One environmental consequence: There has been a resurgence in the population of Rwanda's mountain gorillas, tourism has slowed gorilla habitat loss through deforestation, national park size has been maintained, gorillas are now protected by law, poaching of gorillas by local communities has decreased.

3b Teacher to mark this question.

Sea Ice Express

1a Responses to this question should break down (three) characteristics of the Arctic environment from those listed. Reference should be made to evidence contained in the resources.

2a i The area of sea ice coverage is now limited to the Arctic region north of 80°N (Resources 9A and 9H).

ii Temporal patterns may refer to seasonal variations in sea ice cover (Resource 9I) or variations over a longer period of time (Resource 9H). For example, ice of all ages has declined; 5+-year-old ice has declined quite sharply. Much of the Arctic ice cover now consists of first-year ice (shown in purple), which tends to melt rapidly in summer's warmth.

2b The Arctic Circle is located in the Arctic Ocean and is bordered by eight nations: Greenland, Russia, Norway, USA, Canada, Iceland, Sweden and Finland (Resource 9A).

3a The Arctic plays an important role in climate change: decreasing Arctic sea ice decreases albedo (reflected sunlight), increasing global warming. This potentially leads to rising temperatures, which increase the rate of glacier melting, causing sea levels to rise.

3b Teacher to mark this question. Responses to this question are required to give a detailed account including reasons and/or causes as to how certain elements of the environment interact as specified in a range of resources.

Turbines in Turitea

1a i The generation of electricity increased at a constant rate from 1975 to 2016 (Resource 10A).

ii Responses to this question should identify at least one trend relating to a named energy source from Resource 10A. For example, the generation of geothermal energy was constant between 1975 and 1988 after which there was a steady increase; generation of geothermal energy began to increase at a faster rate in 2006.

1b

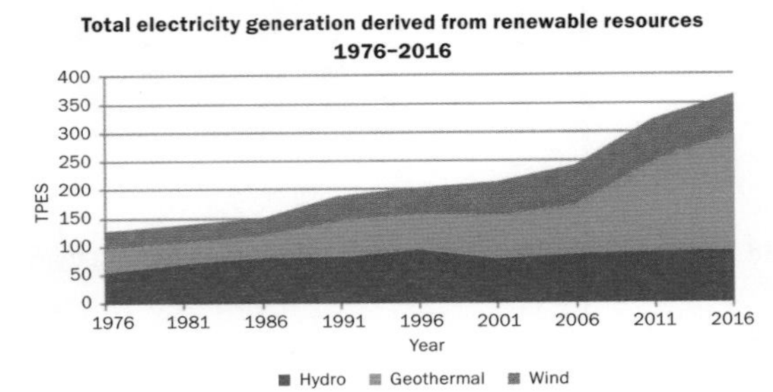

1c New Zealand's demand for electricity increased steadily between 1975 and 2007. However, the last eight years saw consumption plateau.

1d Possible reasons include: a slowdown in population growth relative to past growth, more efficient uses of electricity, price increases in the cost of electricity is causing people reduce their dependence.

1e i Industrial consumption for electricity increased between 1975 and 1991. Since the early 1990s the rate of increase has been minimal. Some quantification of consumption is required (Resource 10C).

ii Residential consumption for electricity increased between 1975 and 2011. Some quantification of consumption is required (Resource 10C). However, in the period since 2011, residential electricity consumption has plateaued.

2a Teacher to mark this question. Ensure all mapping conventions are satisfied (i.e. title, use of key, north point).

2b South-southeast

2c Responses should acknowledge the farm's rural setting albeit in close proximity to nearby urban settlements. For example, the site for Turitea is approximately 10 km south-east of Palmerston North primarily along a 14 km ridge in the northern Tararua Range.

2d i Responses to this question should break down and interpret data or information about one characteristic of the natural environment listed in Resource 10G, e.g. drainage, relief or (native) vegetation.

ii Responses to this question should break down and interpret data or information about one characteristic of the cultural environment listed in Resource 10G, e.g. settlement, rural land use, transportation or (exotic) vegetation.

3a Teacher to mark this question.

3b Teacher to mark this question.

ISBN: 9780170425285